PENGUIN BOOKS

THE PHYSICS OF THE BUFFYVERSE

JENNIFER OUELLETTE is the author of *Black Bodies and Quantum Cats: Tales from the Annals of Physics*. A member of the National Association of Science Writers and Author's Guild, she is associate editor of *APS News*, the monthly membership publication of the American Physical Society. Her work has also appeared in *Discover, Salon*, and *New Scientist*, among other venues, and she maintains a blog called Cocktail Party Physics. She has covered such varied topics as the acoustics of Mayan pyramids and New York City subways; fractal patterns in the paintings of Jackson Pollock; and the precarious pitfalls of pseudoscience. Her article on concert hall acoustics for *The Industrial Physicist* garnered an award in science writing from the Acoustical Society of America. An avid fan of *Buffy the Vampire Slayer* since the series premiered in 1997, she holds a black belt in jujitsu and lives in Washington, DC. Visit her on the Web: www.jenniferouellette-writes.com and www.twistedphysics.typepad.com/cocktail_party_physics.

ABOUT THE ILLUSTRATOR

PAUL DLUGOKENCKY has created cartoons and illustrations for Viking Penguin, King Features Syndicate, *Newsday, The Economic Times* of Long Island, Community Newspapers of Long Island, the American Physical Society, and the American Institute of Physics. He's also the founder and brewer at The Blind Bat Brewery. He lives in Centerport, New York. On the Web: www.aDailyCartoon.com and www.BlindBatBrewery.com.

PENGUIN BOOKS

ILLUSTRATIONS BY

PAUL DLUGOKENCKY

THE PHYSICS OF THE BUFFYVERSE

JENNIFER OUELLETTE

PENGUIN BOOKS
Published by Penguin Group
Penguin Group (USA) Inc., 375 Hudson Street, New York, New York 10014, U.S.A.
Penguin Group (Canada), 90 Eglinton Avenue East, Suite 700, Toronto, Ontario, Canada M4P 2Y3 (a division of Pearson Penguin Canada Inc.)
Penguin Books Ltd, 80 Strand, London WC2R 0RL, England
Penguin Ireland, 25 St Stephen's Green, Dublin 2, Ireland (a division of Penguin Books Ltd)
Penguin Group (Australia), 250 Camberwell Road, Camberwell, Victoria 3124, Australia (a division of Pearson Australia Group Pty Ltd)
Penguin Books India Pvt Ltd, 11 Community Centre, Panchsheel Park, New Delhi – 110 017, India
Penguin Group (NZ), cnr Airborne and Rosedale Roads, Albany, Auckland 1310, New Zealand (a division of Pearson New Zealand Ltd)
Penguin Books (South Africa) (Pty) Ltd, 24 Sturdee Avenue, Rosebank, Johannesburg 2196, South Africa

Penguin Books Ltd, Registered Offices:
80 Strand, London WC2R 0RL, England

First published in Penguin Books 2006

1 3 5 7 9 10 8 6 4 2

Illustrations by Paul Dlugokencky

THE LIBRARY OF CONGRESS CATALOGING IN PUBLICATION DATA
Ouellette, Jennifer.
The physics of the Buffyverse / by Jennifer Ouellette ;
illustrations by Paul Dlugokencky.
p. cm.
Includes bibliographical references and index.
ISBN 978-0-14-303862-7
1. Physics—Micellanea. 2. Physics—Humor. 3. Buffy, the vampire slayer (Television program) I. Title.
QC75.084 2007
530—dc22 2006025634

Printed in the United States of America
Set in Minion and Trade Gothic
Designed by Judith Stagnitto Abbate / Abbate Design

For the fandom

Contents

"Science must begin with myths."
—SIR KARL POPPER

INTRODUCTION

Welcome to the Buffyverse

"Hell's empty, and all the demons are here."
—Ariel, THE TEMPEST

It begins with the sound of shattering glass. A young man and his pretty blond date break into the science lab at the local high school late one night for a bit of mischief—most likely to engage in some extracurricular hanky-panky on the roof. The girl appears nervous, starting at every sound, fearful that someone, or something, with evil intentions, is lurking in the darkened school. The young man has all the arrogance of youth, dismissing her fears and assuring her with an insinuating leer that they are quite alone. Whereupon the girl's face transforms into that of a fanged, yellow-eyed demon, and she sinks her teeth into her soon-to-be-former date's neck.

This is the weird yet wonderful world of the Buffyverse, where magic, vampires, and demons are real, and mystical convergences and otherworldly phenomena are everyday occurrences. When *Buffy the Vampire Slayer* debuted as a midseason replacement in 1997, few industry insiders expected it to do well. After all, the campy film version had tanked at the box office. Actor Kiefer Sutherland—whose father, Donald Sutherland, co-starred in the film—reportedly was so pessimistic about its chances that he told the show's star, Sarah Michelle Gellar, not to worry, because she was bound to get another series later on. But the TV show defied the naysayers and ended up running for seven seasons. While it never achieved the blockbuster popularity of mainstream sitcoms like *Friends* or *Seinfeld, Buffy* quickly attracted a strong cult following, drawn by its unique blend of horror, science fiction, and high school melodrama. The show also became a critics' darling, thanks to generous sprinklings of mythology, literary allusion, biting wit, and a lexicon of its own hip teen lingo (dubbed "Buffyspeak").

The premise is simple enough: "Into every generation, a Slayer is born, one girl with the strength and skill to hunt the vampires." That girl is fifteen-year-old Buffy Summers. In the pilot episode ("Welcome to the Hellmouth"), Buffy moves to the fictional town of Sunnydale, California, with her divorced mother, Joyce, after Buffy is expelled from her former high school in Los Angeles. (She burned down the gym, but there were extenuating circumstances: It was full of vampires.)

Sunnydale is not the picture-perfect town that it seems to be on the surface. It is located squarely on top of a Hellmouth, a mystical portal between the world of Sunnydale and a separate hell dimension. The Hellmouth emits all kinds of bad juju, and its energy draws evil beings to the area like a giant magnet of badness. Buffy's job is to keep the demons at bay and prevent hell from erupting on Earth. She does so for the next seven years, beating back everything from vampires to hell gods to the very First

Evil, while simultaneously grappling with the usual travails of high school, college, and the onset of young adulthood—all of which can be scarier than any demon horde.

Fortunately, she doesn't fight alone. Buffy is aided by her oh-so-British Watcher, Rupert Giles, and her new friends: Willow, Xander, and Angel—a reformed vampire cursed by gypsies who restored his human soul. In 1999, Angel became the star of his very own eponymous spinoff series (*Angel*). He sets up shop as a private investigator to fight injustice and help the hopeless in a fictionalized version of Los Angeles—which usually involves killing demons and battling other forces of evil. The characters and events that populate these two series make up what is known as the Buffyverse.

On the surface this surreal, fictional world would appear to have very little to do with the world of science. Science, especially physics, views the universe as a gigantic, complex machine that operates in accordance with a handful of underlying fundamental principles: the laws of physics. Magic and superstition rightly have no place in serious science. Tell a physicist that you're interested in exploring the physics of the Buffyverse, and the most likely response will be a blank, puzzled stare, followed by a dubious observation: "But vampires aren't real. . . ." The skepticism is understandable. But look a bit closer, and you'll find that science lurks everywhere in the Buffyverse, from the "Big Picture" framework to the nooks and crannies. It's not just relegated to Sunnydale High School's science lab.

For instance, many of the monsters' traits are drawn from real-world biology, such as demons that inject their victims with poisonous toxins to paralyze them before they feed. Vampirism could be viewed as an infectious disease, spreading through contamination of the blood, almost a modern metaphor for AIDS. The ancient demon Illyria reemerges from a multimillennium-long sleep in "A Hole in the World" (*Angel*, Season 5, or, henceforward, A-5) as a form of biological warfare. Just like a virus, she infects her host, killing that host so that she can inhabit the shell that remains. The host becomes a potential weapon of mass destruction. Any attempt to extract Illyria from her victim would make the virus "airborne"; thousands would die, instead of just one person.

Chemistry is plainly evident in the concoction of brews and potions for the casting of spells. In "Witch" (*Buffy*, Season 1, or, henceforward, B-1), Xander and Willow make use of the ingredients in their science class to concoct a potion that will tell them if their classmate Amy is a witch—although they have to improvise a bit, obtaining the "eye of newt" during their dissection of a frog. When Buffy's mother becomes mysteriously ill ("No Place Like Home," B-5), Buffy suspects that it might be the result of a magic spell. She performs her own spell called *tirer la couture*—literally, "pull the curtain back" in French, although Buffy (who didn't do well in French class) mistranslates it as "rotate many foodstuffs." All spells leave a trace signature normally invisible to humans, and her spell enables Buffy to see these traces to determine whether a spell has been cast. The concept is very similar to chemical elements' having distinct "signatures," in the form of emitted light (electromagnetic radiation) that is undetectable to human eyes. We can detect this light with instruments called spectrometers. The color of the light tells us which elements are present in a given sample, while the intensity of that color indicates how much of a particular element is present.

As for physics, writers for both series have openly drawn on specific concepts in quantum mechanics, relativity, and string theory to develop innovative plots for episodes. A high school girl becomes invisible after months of nobody noticing her—a clever twist on the quantum notion that observation determines the outcome of a subatomic-scale experiment ("Out of Mind, Out of

Sight," B-1). There are teleporting demons, temporal folds, time loops, and dimensional portals, conceptually similar to the hypothetical wormholes proposed by real-world physicists. And one critical scene in an *Angel* episode takes place at a scientific symposium on string theory ("Supersymmetry," A-4). The Buffyverse has seeped into physics in turn. In December 2005, astronomers found that a small object in a ring of icy bodies near Neptune (known as the Kuiper belt) had an unusually tilted orbit. They dubbed the object "Buffy," in part because—like many things in the Buffyverse—its orbit can't be explained by the prevailing scientific theories of how the outer solar system formed.

More generally, Buffy and her entire gang of "Scoobies"—a reference to those meddling kids in the cartoon *Scooby-Doo*—know the value of doing their homework. When some new evil comes to town, the first thing they do is launch into "research mode." Angel and his team of fellow demon hunters adopt the same approach. Skipping that vital step is usually a recipe for failure. In the same way that scientists must first understand the nature of a problem before they can design successful theories and experiments, the Scoobies and "Team Angel" understand that they must first understand the nature of the thing they are fighting in order to defeat it.

There are technological parallels as well. The books in the library of Wolfram & Hart (aka "the devil's law firm") on *Angel* are blank until someone asks for a specific tome. Then the pages fill with the requested text. Electronic paper is a similar real-world technology that is already being used for commercial signage in the marketplace. In "Witch" (B-1), Buffy uses a mirror to reflect the energy of a witch's spell back onto the witch. The technique is similar in concept to Alexander Graham Bell's photophone, an early forerunner to fiber optic communication. The photophone transmitted sound on a beam of light to a mirror, causing the mirror to vibrate in response. The instrument then captured the vibrations that reflected off the mirror and transformed them back into sound.

Even the most familiar technology gets a new twist. The demon puppets in "Smile Time" (A-5) use the TV signal of their hit children's show as a two-way conduit. They graft a hidden carrier signal onto the regular broadcast signal—camouflaged by a magic spell—that enables them to communicate individually with their young viewers and sap their innocence away. In "I Robot, You Jane" (B-1) a demon who has been bound into an ancient mystical book goes binary, unleashed on the Internet when Willow scans the text into a computer. The demon's essence is broken into electron "bits," much like radio and TV signals, and then digitized into the "bytes" used in computers. Giles and the school's computer science teacher, Jenny Calendar, must combine magic with information technology to defeat the demon: They form a virtual mystical circle in an online chat room to cast a "rebinding" spell.

This melding of magic and science is a defining feature of the Buffyverse. *Buffy* and *Angel* creator Joss Whedon has said that the original series was intended as a metaphor for how high school can sometimes seem like hell to teenagers. He made his fictional high school a literal hell, with vampires and other monsters embodying humanity's inner demons. The same can be said for the physics in the series. Sometimes it takes center stage, but more often than not, it's woven into the fabric of the fictive framework, and works best on a metaphorical level. The Buffyverse is ruled largely by metaphysics. Try to interpret things too literally, and one quickly runs into absurdities, much the same way that attempting to precisely determine two mutually exclusive properties of a subatomic particle leads to unwanted mathematical "singularities."

That doesn't mean it has nothing to teach us. Metaphor can

be a powerful tool for communicating abstract concepts, and the Buffyverse offers a unique lens through which to view the precepts of science with fresh eyes. It may be filled with creatures and events that defy our laws of physics, but Whedon's world has its own versions of these laws. Gravity and electromagnetism behave much like the corresponding forces in our own universe. Entropy is also present: The shattered windowpane in the science lab doesn't spontaneously piece itself back together. Then there is conservation of energy: you can't get something from nothing. Magic is never the "easy" alternative, even if it seems that way. There are costs incurred, consequences that must be paid, and above all, built-in restrictions as to how and when it can be used. "The universe has rules," Winifred ("Fred") Berkel opines in "Heartthrob" (A-3) after Angel, being a vampire, must ask permission before entering her room.

Talented magical practitioners, like *Buffy*'s Willow, understand these rules, and learn (sooner or later) to work within the limitations they impose. The magic demands it. Physicists do the same in the real world. Mathematics is the language of physics, the key to unlocking the secrets of our universe, whereas mystical languages in the Buffyverse have the power to cast spells, open portals, or otherwise alter reality. Yet there is a difference between what is mathematically possible and what is practically possible in physics, just as there is a difference between what Willow dreams of accomplishing with her magic and what she can actually achieve.

The key difference between our world and Buffy's is that in her world, there is a mysterious "X" factor: a seemingly unlimited reservoir of extra mystical energy that allows for phenomena that simply can't happen in our universe, at least not with our present capabilities. Once that fictional factor is figured into the equation, the Buffyverse is impressively consistent in following its own internal rules. It's not an entirely absurd premise. Physicists believe that our own universe has a mysterious form of "dark energy" that accounts for its accelerating expansion rate, and that empty space is brimming with virtual particles that can produce fleeting, minuscule amounts of "negative energy." In fact, much of the science and technology we take for granted today would be considered magic by someone living just 150 years ago—a point made by the famed science-fiction author Arthur C. Clarke way back in 1961.* In that sense, "magic" is just a term to describe phenomena we don't yet fully comprehend.

Nonetheless, there are inconsistencies in the Buffyverse, and fans are often the first to point them out. Many a discussion in an online forum has become mired in a recitation of internal contradictions. For instance, Angel can't perform CPR on Buffy when she is briefly dead from drowning because vampires don't breathe, yet his former vampire protégé, Spike, seems to have no problem blowing cigarette smoke through his nostrils. Vampires supposedly don't have circulation, and yet they bleed when injured. They don't cast a reflection, yet they show up on camera. Buffy bends the metal barrel of a rifle in "Phases" (B-2) yet can't break free when she's chained up in "Crush" (B-5). The list goes on and on, prompting one anonymous newsgroup poster to declare that while a book had been written on *The Physics of Star Trek* (by physicist Lawrence Krauss), "I don't think there will ever be a book called *The Physics of the Buffyverse*." As one would say in Buffyspeak, soak in the irony for a moment.

It's easy to get so hung up on the inconsistencies—which plague every sci-fi series, including *Star Trek*—that one loses sight of the primary objective: to entertain. *Buffy* and *Angel* were

*In *Profiles of the Future*, Clarke elucidated his Third Law: "Any sufficiently advanced technology is indistinguishable from magic."

never intended to be documentaries, and it hardly seems fair to hold them to the same exacting high standard of accuracy. Far better to adopt the irreverent approach of the writers themselves, who occasionally make gleeful in-jokes at their own expense, and poke fun at the contradictions. More often than not, they get it right—or at least they get it half right. Like all good science-fiction writers, they make creative extrapolations, taking an idea and playing with potential implications.

One naturally wonders how much of this was a deliberate attempt to incorporate physics principles into the Buffyverse and how much is merely coincidence. Yet ultimately this question is irrelevant, since it ignores the complex nature of the creative process. There are cases where the connection is deliberate, such as the direct references to energy conservation, quantum entanglement, black holes, and string theory. Sometimes the writers put their own unique stamp on the stock conventions of the sci-fi genre—teleportation, portals to other worlds, or time loops—that are firmly rooted in theoretical physics. Sometimes it comes about because physics is so intrinsic to the daily conveniences we take for granted: TVs, radios, microwaves, refrigerators, electronic security systems, computers, and cell phones—the last of which a frustrated Angel insists were "cooked up by a bored warlock" as he struggles to find a decent signal while tracking a demon in his convertible ("She," A-1). And sometimes it's just that creating a believable fictional universe requires incorporating certain recognizable attributes of our own.

Writers themselves aren't always sure of their intentions. They follow their creative instincts in what is often a continually unfolding process of discovery. Whedon has admitted in interviews that he didn't really work out the physics of the Buffyverse extensively beforehand, and bluffed quite a bit of it as he went along. His fictional physical laws clearly evolve along with the characters and plot lines as the two series progress. Small wonder, then, that there are inconsistencies. But in some sense, this makes his world seem more, not less, real to the viewer. After all, there's a great deal we still don't understand about our own universe, and scientists in ages past used to believe some very strange things. The physical laws themselves may not have changed, but our understanding of them, like the Buffyverse, has evolved over time. Physics progresses through trial and error, building on all the knowledge that came before. Isaac Newton is considered one of the greatest scientists who ever lived, yet he memorably claimed that he only saw farther because he "stood on the shoulders of giants."

Of course—to baldly state the obvious—one shouldn't confuse science with science fiction. There are elements in the Buffyverse with striking parallels to real-world physics, but it is still a make-believe universe (or, more accurately, a multiverse). Unlike real-world physics, the creators can (and sometimes do) opt to "change" these laws to suit their narrative needs. The point of this book is not to provide a rational, scientific explanation for every single strange occurrence in Whedon's world. That would be an exercise in futility—and wouldn't be much fun. There are definitely aspects that can only be attributed to that helpful catchphrase "artistic license," or excused by the viewer's willing suspension of disbelief. We can still indulge in a little playful conjecture, in keeping with the spirit of the series, even if the underlying science is profoundly serious.

No doubt there are those who will find this lighthearted, irreverent approach to explicating physics discomfiting; some people prefer their physics straight, with no pop-culture chaser. That's a matter of personal taste. But just because something is populist in tone doesn't automatically mean it lacks substance. Buffy's long-suffering Watcher, Giles, initially bemoans the frivolous (to him) pursuits of his spirited young charge. After all,

Buffy insists on juggling her Slayer duties with dates, dances, hanging out at the Bronze, girl talks with Willow over nonfat lattes, and frequent excursions to Sunnydale's shopping mall. Yet Giles soon realizes that her arch, airhead demeanor masks genuine intelligence and seriousness of purpose, while her unconventional approach yields novel solutions to seemingly intractable problems—a most scientific objective.

Buffy begins in a science lab, with a shattered windowpane and a cocky young man whose entire world is about to be turned on its head. Physics brings us that same element of surprise. Just when we're starting to feel comfortable and complacent with what we know about our universe, and our place in it, we are blindsided by something unexpected that shatters our preconceptions. The earth is round, not flat. The sun, and not the earth, is at the center of the solar system. Time is not absolute. The expansion of the universe is speeding up instead of slowing down. Or we come to realize (much too late) that vampires exist and humans are not, after all, at the top of the food chain.

Welcome to the Buffyverse.

WELCOME TO THE BUFFYVERSE
E=MC²
WELCOME TO
SUNNYDALE
Enjoy Your Stay!

1

CREATURE FEATURE

Physics Meets Biology in a Monster Mash

"Art, like nature, has her monsters,
things of bestial shape and with hideous voices."
—OSCAR WILDE

One starry night, Willow and her girlfriend, Tara, are gazing at the constellations when they spot a meteor streaking across the sky. It's not an ordinary meteor: this one has a soft, chewy demon center, unleashing an overgrown, slimy, lizardlike creature onto a community already overrun with demons. The creature vomits a sticky, odiferous substance onto its victim's face, which then hardens, suffocating said victim. Xander dubs it a killer snot monster, but the Scoobies soon learn that it is called a Queller demon.

The Queller demon featured in "Listening to Fear" is an unusual departure from the Buffyverse's monstrous menagerie. Most of its cornucopia of mystical creatures don't hail from outer space, but are the remnant of an earlier mythological era when demons ruled the Earth, before evacuating to alternative hell dimensions. Those that remain are hybrids, a mix of biological and demonic elements. And while they are entirely fictional, many of those creatures call to mind attributes of real-world creatures found in nature.

For instance, the She-Mantis in "Teacher's Pet" (B-1) behaves like a real praying mantis: she can turn her head a full 180 degrees, eats crickets sandwiched between slices of Wonder Bread for lunch, lays eggs and fertilizes them via young male virgins, then bites off the heads of her mates afterward. The "Inca Mummy Girl" (B-2) derives sustenance by sucking out the vital juices from other humans, leaving a dried-up shell behind, much like spiders do with their prey. And in "Selfless" (B-7), Buffy slays a large spiderlike demon that produces a sticky, black webbing. Even the extraterrestrial Queller demon has an earthly biological counterpart: the hagfish, an eel-like creature that, when attacked, excretes large amounts of mucus that can ensnare and suffocate the target. The hagfish is nature's killer snot monster.

The various types of demons in the Buffyverse are so numerous that Team Angel resorts to simply listing the monsters of the week as NDUOs—Nasty Demons of Unknown Origin—until the species has been identified ("Judgment," A-2). Vampires are just the tip of the iceberg; in fact, they're frequently dismissed as "blood rats" by other demons. Yet they are among the most complex species of monster. They have an extensive mythological lineage, for one thing, as well as some fascinating physiological attributes that straddle the interface among physics, biology, and chemistry. The challenge, when it comes to vampires, is to separate the few shreds of fact from the sticky morass of fiction.

DEAD MEN WALKING

Buffy's routine patrol of Sunnydale's cemeteries turns extraordinary when she encounters a tall, dark, and handsome vampire with a lilting Eastern European accent. He claims to be none other than the legendary Count Dracula ("Buffy vs. Dracula," B-5). The Scoobies quickly discover that Dracula has special abilities: He can morph into other creatures (a bat, or a wolf) or disappear in a puff of smoke, and his intense gaze is literally hypnotic.

Spike rightly dismisses these powers as "gypsy parlor tricks." Dracula is merely an accomplished mystical illusionist; in all other respects, he is a garden-variety vampire. Hypnosis, or mesmerism, as it was also called, was debunked in the eighteenth century by a panel of scientists that included American founding father Benjamin Franklin, and more recently on cable TV by that morbidly magical duo Penn and Teller. It has more to do with the power of suggestion and human gullibility than with any special powers. Changing a human figure into a bat, or into smoke, are common illusions that have been performed by magicians for centuries. Even assuming that Dracula were able to change himself into a bat, changing back would be much harder, thanks to the inherent limitations imposed by mass/energy conservation and entropy.* His greatest power appears to be his celebrity status: The star-struck Scoobies are each dazzled in turn, until Buffy manages to shake off the count's thrall long enough to stake him . . . twice. (Being Dracula, he always comes back.)

Dracula might be the most famous bloodsucker, much to

*These concepts are discussed in detail in chapter 3.

Spike's chagrin, but tales of vampirelike creatures date back at least four thousand years to ancient Mesopotamia. The Assyrians feared a demon goddess called Lamastu (literally, "she who erases"), who killed babies in their cribs, or while still in the womb. Ancient Jewish texts mention a similar creature, Lilith, Adam's first wife, who steals away infants and unborn children. Neither of these could be considered "vampires" in the modern sense, but they are the precursors to the Greek legend of Lamia, an immortal monster who sucked the blood from young children.

In Chinese folklore another type of protovampire, called the *k'uei,* were reanimated corpses that rose from the grave and preyed on the living. One type, the *Kuang-shi* (or *Chiang-shi*), could fly and take different forms, biting into its prey with sharp fangs. The Russian *upir,* Indian *vetala*, and Greek *vrykolakas* were also undead corpses that rose from their graves to feed off the living. In the early Middle Ages, Russian villagers would exhume suspect corpses and destroy the body by cremation, decapitation, or by driving a wooden stake through the heart. Stakes were often secured above corpses upon burial, so the creature would impale itself if it tried to escape. One wonders why the people of Sunnydale didn't think of that.

By the nineteenth century, most of Europe was consumed by vampire hysteria, eventually inspiring a theater manager in London named Bram Stoker to pen the novel *Dracula*. Stoker based his villain on a Wallachian prince from the mid-1400s, Vlad Dracula (or "Vlad the Impaler"), although he wasn't a vampire—just excessively violent, in a particularly violent age. He liked to impale his enemies on long wooden spikes. Another likely influence on Stoker's creation was Elizabeth Bathory, a Transylvanian noblewoman of the late sixteenth century and one of the few historically documented "vampires," although she, too, was quite mortal. Her greatest power was the protection from prosecution afforded by her aristocratic lineage. Known as "the bloody countess," she practiced her own sadistic form of homeopathic alternative medicine: She bathed in the blood of young women (more than six hundred victims in all) to preserve her youthful appearance, and was rumored to drink the blood of especially pretty ones. Her heinous practices were far from effective: She continued to age, and died in 1614 in her early fifties while imprisoned for her crimes in her own castle, deep in the Carpathian Mountains.

Drawing liberally on this rich tradition, the Buffyverse boasts its own Undead mythology. We learn in "The Harvest" (B-1) that the first vampire was created when the last demon to leave the Earth mixed its blood with a human. As Buffy explains in her inimitable style, "To make you a vampire, they have to suck your blood. And then you have to suck their blood. It's like a whole big sucking thing." ("Welcome to the Hellmouth," B-1) This icky-sounding process is known as "siring," and it is the only way to turn someone into a vampire; merely feeding off a victim is insufficient. The human dies in the transformation, the soul departs, and a demon sets up shop in the empty vessel. A vampire retains the form and memories of its human vessel, but at heart, it is a demon. Clinically dead, with no pulse or body temperature, a vampire can pass for human until it is ready to feed. Then its face distorts into the fanged visage of the monster inside.

Like the Dracula legend, the vampires in the Buffyverse owe their special powers mostly to mythology and superstition. They possess superhuman strength and rapid healing ability, and must be invited into a private home before they can enter. Buffyverse vampires are also hypersensitive to crosses, sunlight, and holy water, and disintegrate into a thick cloud of ash when staked or decapitated—the only ways they can be killed, apart from fire or direct sunlight, which cause them to burst into flame, a sort of spontaneous vampiric combustion. Few of these features have any basis in real-world science, and there's no reason they should, since we're talking about mythical creatures.

Yet there *are* aspects of vampire lore that resemble the symptoms of real diseases. Most notably, porphyria is a hereditary disease in which the body doesn't produce sufficient heme, an iron-rich pigment in the blood. Those who suffer from certain types of porphyria are highly sensitive to sunlight and may have reddish mouths, like the ancient Master vampire with "fruit-punch mouth" who was introduced in Season 1 of *Buffy*. Some historians suspect that a common folk remedy for porphyria may have been to drink fresh blood, but if so, those efforts were wasted. The chemical enzymes in the blood that sufferers require can't survive the digestive process; they must go straight into the bloodstream via blood transfusions or injections.

Over time the most severe (and rarest) forms of porphyria can cause blistering, scarring and thickening of the skin, and in extreme cases can lead to disfigurement. The lips and gums may become so taut that the teeth protrude like fangs, giving the sufferer an appearance strikingly similar to the Nosferatu of early horror films, or the Buffyverse's Turok-han, an ancient race of *über*vamps. In fact, the writers of *Buffy* are on record as saying that they originally conceived of vampirism as a progressive disease, and the Master's appearance reflected that. But the similarities between vampirism and the symptoms of porphyria appear to be entirely coincidental. There have been only two hundred or so documented cases of the most extreme forms of porphyria, hardly enough to inspire the plethora of vampire legends around the world, and many of the cited vampiric attributes didn't appear in folklore until the nineteenth century.

We can find clues to explain vampires' extreme sensitivity to sunlight not just in the enzyme deficiencies that characterize porphyria and similar disorders, but also by looking at how the sun's rays cause human skin to tan and burn. The sun emits three forms of light: infrared light (heat), visible light, and ultraviolet (UV) light. It is the latter that is responsible for skin damage: Prolonged exposure can damage and kill skin cells, which then release chemicals that activate the body's pain receptors. The reddening of sunburned skin is the result of increased blood flow to the damaged areas in order to remove the dead cells. The energy from UV light also stimulates the production of a pigment known as melanin, which causes the skin to darken, or tan. Melanin actually absorbs the UV radiation in sunlight, protecting skin cells from further damage.

Melanin is produced gradually, which is why would-be tanners must build up levels of the protective pigment in their skin cells over the course of several days. It's also why darker-skinned people are less likely to burn or suffer from skin cancer than those of fairer complexion: They possess naturally high levels of melanin. In contrast, albinos don't have any melanin at all in their skin, hair, or irises, because they are missing a critical enzyme required for its production. This makes them especially vulnerable to the sun's UV rays, although they don't burst into flame. Vampires are not albinos: The Slayer-obsessed vampire, Spike, owes his signature bleached-white hair to a bottle of peroxide, and there are several dark-skinned vampires scattered throughout the Buffyverse, like Mayor Wilkins's henchman, Mr. Trick. Still, one could argue that vampires in the Buffyverse possess much less protective melanin than their human counterparts. Giles even refers to Spike as being "melanin deprived" in "Out of My Mind" (B-5), although that lack is more likely to be an effect of a vampire's need to avoid direct sunlight than the cause of such a creature's extreme photosensitivity.

NIGHT STALKERS

Being predatory, nocturnal creatures that can't abide sunlight, it's not surprising that vampires possess certain innate advantages that enable them to navigate in the dark, most notably excellent night vision. Bats, cats, and certain insects possess similar abilities. When Spike, also known as "William the Bloody," first arrives in Sunnydale in "School Hard" (B-2), intent on killing his third Slayer, he traps Buffy and several others in the high school on Parent-Teacher Night. His first move is to cut the power, and hence all the lights. It gives him and his cadre of vampire goons a distinct advantage over their human prey: the ability to see well in the darkened school.

This isn't necessarily a mystical attribute. There is a perfectly reasonable scientific explanation, based in part on biology—vampires are clearly a unique species and have developed certain evolutionary adaptations—and in part on the nature of light and how it interacts with matter. We are able to see the world around us because light reflects off objects and radiates outward in all directions. Our eyes detect this light, which the brain then forms into recognizable images. At night, there is considerably less light available, so human vision is more limited. Light is a physical phenomenon that appears to behave much the same way in the Buffyverse as in our own world.

What exactly is light, and where does it come from? Scientists refer to light as electromagnetic radiation, because light waves are the product of electric and magnetic fields traveling together through space. On a deeper level, light is built into the very nature of atomic structure. Atoms are constantly in motion, even those that appear to make up a solid object. Every atom has electrons that orbit around the atomic nucleus in well-defined, natural orbits—called orbitals—much like the planets in our solar system orbit the Sun. But unlike planets, an electron isn't limited to a single orbit. It can move to higher and lower orbitals as it absorbs and emits excess energy. So the particular orbital an electron inhabits determines its energy at a given point in time. Atomic energy is "quantized": that is, an electron can only increase or decrease its energy in specific, fixed amounts, much like an ATM machine can only dispense funds in multiples of $20 bills. Those fixed amounts correspond to the energy levels of atomic orbitals. An atom's "ground state" energy level is its normal state, but energy can be added—through the application of heat, light, or electricity, for example—and this will cause an electron to jump to a higher orbital.

It is merely a temporary ascension, however. Eventually the electron will want to return to its ground state. To do so, it must get rid of the extra energy, which it emits in discrete packets of energy (the equivalent of the ATM's $20 bills) called photons—particles of light. To a scientist, "light" is a broad term that incorporates not just the light we can see, but also radio waves, microwaves, the sun's UV rays, X-rays, and gamma rays—all of which have different frequencies and wavelengths, and form the full electromagnetic spectrum. The "color" (or wavelength) of the photon emitted depends on its frequency, and this in turn depends on how far the electron had to drop to reach its ground state. If an electron drops only one orbital level, there is less extra energy to dispose of in order to return to its ground state. So the electron will emit a photon with low frequency and a longer wavelength, such as a radio wave. But if an electron drops several orbitals, there is a great deal of excess energy. The electron must emit a photon with higher frequency and shorter wavelengths, such as X-rays. Visible light falls between infrared (IR) and UV light in the electromagnetic spectrum.

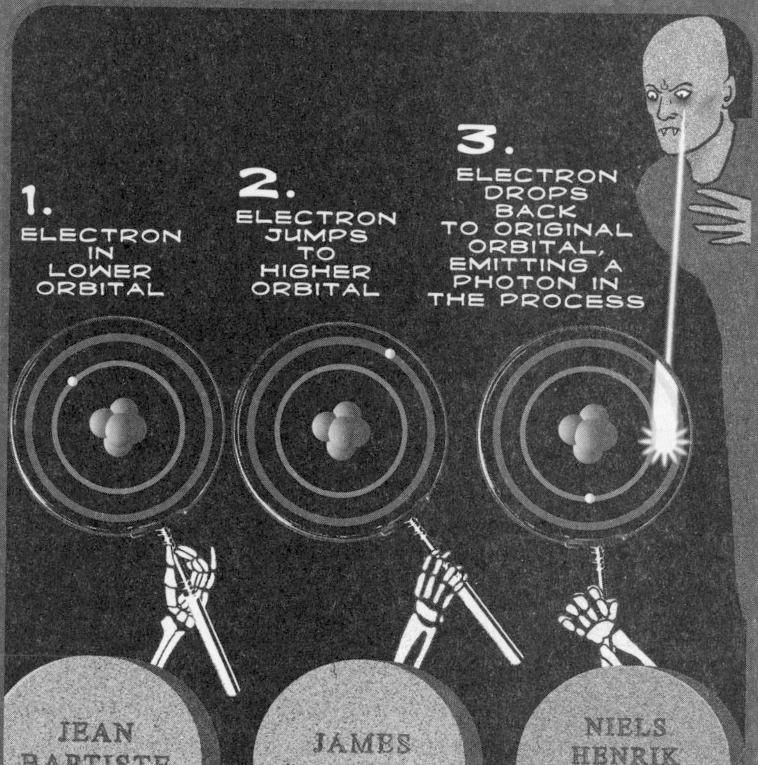
1.
ELECTRON IN LOWER ORBITAL
2.
ELECTRON JUMPS TO HIGHER ORBITAL
3.
ELECTRON DROPS BACK TO ORIGINAL ORBITAL, EMITTING A PHOTON IN THE PROCESS
JEAN BAPTISTE PERRIN
SEPT 30, 1870
APRIL 17, 1942
JAMES CHADWICK
OCT 20, 1897
JULY 24, 1959
NIELS HENRIK DAVID BOHR
OCT 7, 1885

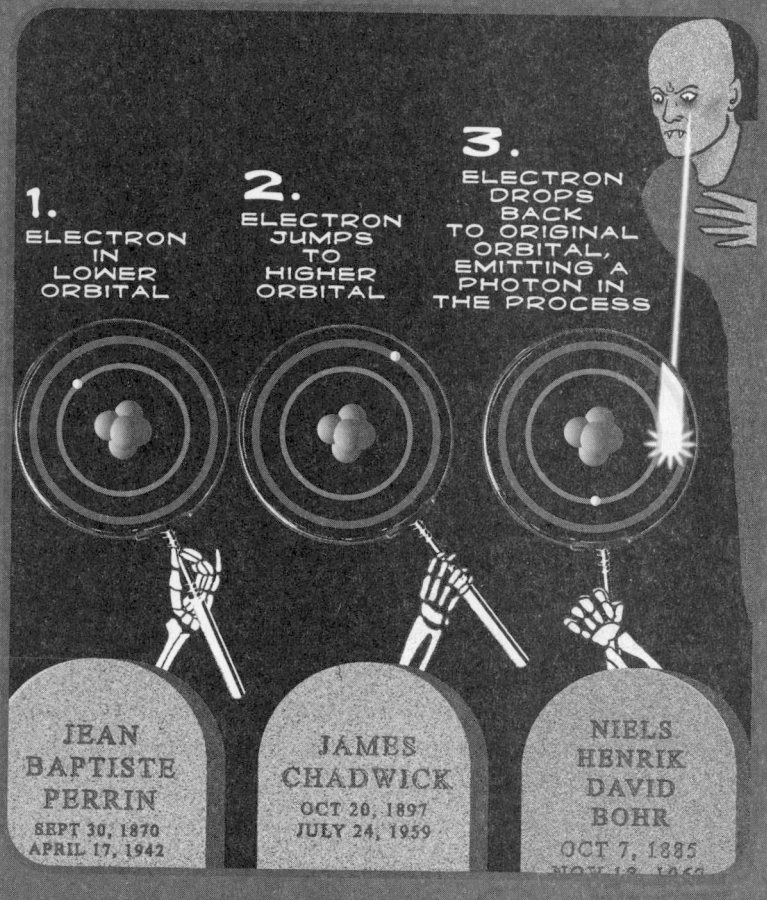

Given the fact that there is very little visible light available to reflect off of objects at night, how, exactly, are vampires "seeing"? There are two possibilities: adaptations in the eye that intensify reflected light, enabling the creature to see even in the dimmest of light, or specialized organs capable of detecting heat emissions from potential prey.

The human eye works in much the same way as a camera captures images on film. The eye's "film" is the retina—a thin layer of neural tissue lining the back of the eye. The tissue is made of photoreceptor cells that receive light and other cells that interpret this information and send the signal to the brain via the optic nerve. The retina contains two kinds of photoreceptor cells: cone cells and rod cells. Cone cells are sensitive to bright light, and can perceive colors. Visible light is made up of a spectrum of colors: red, orange, yellow, green, blue, indigo, and violet (hence the mnemonic Roy G Biv). The human eye has three types of cone cells, each sensitive to a particular primary color of light: blue, green, or red. These three primary colors can mix in the eye so we can see more-complex shades, such as violet or orange. Objects absorb some colors of light and reflect others, and this determines the colors that we see. When light hits a bright red apple, for instance, the apple's surface absorbs all the wavelengths except red, which is reflected to the eye. So we perceive the apple as being red.

In contrast, rod cells work best in low light, and can perceive black-and-white images. A chemical called rhodopsin, found in the rod cells of the eye, is the key to good night vision. Whenever a rhodopsin molecule absorbs a photon, it splits into two secondary molecules—retinal and opsin—that later recombine back into rhodopsin. That's why most of us need a few minutes for our eyes to adjust to the dark when the lights go out. Exposure to bright light causes the rhodopsin to break down, impeding vision, until the two molecules combine back into rhodopsin. Vampires may have much higher levels of rhodopsin, not to mention more rod cells, than the average human, so their night vision is significantly better. It also means they would have fewer cone cells and thus wouldn't perceive colors as well as humans, since most mammals with good night vision are at least partially color blind. This might explain why vampires are attracted to bright colors ("The Wish," B-3): they find them easier to see than more muted tones.

That doesn't mean that vampires can see in pitch-darkness. Even Spike finds himself flicking on his lighter to see the crude drawings on the walls of a remote cave that houses a demon with the power to restore his soul ("Villains," B-6). At least a smidgen of light is needed to reflect off objects so vampires can see—even if it's just starlight or moonlight. Cats are also nocturnal hunters, but they can't see in pitch-black conditions either, despite having more rod cells (and therefore, rhodopsin) than humans. Their keen night vision is due to a special layer of cells lining the back of the feline retina, called the *tapetum lucidum*. It acts like a mirror, collecting and reflecting light through the rods a second time to restimulate them. The result is a double exposure of light, permitting cats to see in near-darkness. When we see an eerie yellow glow in a cat's eyes in flash photographs, we are looking at the light reflecting off the *tapetum lucidum*. Perhaps it is no accident that a vampire's demonic visage features glowing yellow eyes.

Another possibility for vampiric night vision is heat. Heat is a form of light, technically known as thermal infrared radiation. Thermal radiation falls just below the range of visible light on the electromagnetic spectrum. It's the reason a stovetop burner glows red: The atoms in the burner are excited by the influx of energy when the burner turns on, causing the emission of photons in that region of the electromagnetic spectrum. All objects radiate heat. Living things emit more heat than, say, a rock, because they must consume energy to stay alive, and this in turn generates heat:

The energized atoms emit photons in the thermal IR spectrum as they return to their natural ground state.

This is a boon to the Initiative, a secret, demon-hunting government agency based in Sunnydale. Using infrared detectors, its operatives are able to locate a vampire moving among humans. Vampires, being dead, have a lower thermal output and thus show up at exactly the temperature of their environment, compared to a human's 98.6 degrees Fahrenheit. So if one person in a crowded room registers at room temperature, compared to surrounding hotter bodies, that person is most likely a vampire. "We got a cold one," one soldier says upon locating Spike with a similar device ("The Initiative," B-4). Modern night-vision equipment exploits the generation of heat from living bodies by focusing the thermal emissions with a special lens, then transmitting it to IR detectors, which create a detailed pattern based on variations in temperature. This pattern is called a thermogram. The thermogram is then translated into electrical impulses, and these are analyzed by a computer and sent to the display, which, depending on the intensity of the IR emission, shows the data as various colors. But vampires don't need night-vision goggles or detectors, so they would have to be naturally more sensitive to heat.

Despite their human origins, vampires are vaguely reptilian in appearance when they don their demonic visage. Perhaps not coincidentally, certain species of snakes rely on detecting thermal emissions to hunt and capture their prey, the most notable example being the pit viper. Yet another night stalker, the pit viper gets its name from two thermoreceptors that resemble small pits located on either side of its head. These serve as sensors to determine both the amount of heat and the distance to the heat source, because they contain membranes that are sensitive to infrared radiation. Pit vipers can detect any prey whose body temperature is hotter or colder than the surrounding environment—even if the temperature difference is merely fractions of a degree. Scientists are still debating the degree to which snakes like the pit viper merely feel heat, and how much of this information is processed as an image by the reptile's brain. But the human brain is much more complex than a reptile's, and vampires were once human. So if they possessed a similar heat-sensing ability, they would most likely be able to process the equivalent of thermograms: That is what they would see.

Vampires also have their own form of camouflage: They don't cast a reflection, which enables them to sneak up on potential prey from behind. When Buffy and Angel are sitting side by side on her bed one night talking ("What's My Line?" B-2), Buffy looks over at the vanity mirror and sees only herself reflected. Many creatures in nature employ some sort of camouflage, blending so well with their background environment that they are virtually undetectable by predators or their own prey. Octopuses, for example, can change color and shape. The praying mantis can mimic leaves, twigs, and flowers.

Chameleons are the best-known example of a creature's ability to change its color. They can do this because they have several layers of cells containing pigment granules that are controlled by the central nervous system. Under the transparent outer skin are two cell layers containing red and yellow pigments; under those are cell layers that reflect blue and white light, supported by an even deeper layer of brown melanin. Which colors are prevalent depends on the creature's mood and environment. Many people assume that chameleons change color in response to their surroundings, but this is not quite correct. While the ability does help the animal hide, its primary purpose appears to be the regulation of body temperature. Dark colors absorb heat, and light colors repel it, so chameleons darken when they are cold, for example. They also display different colors when mating or when defending their territory.

While not precisely camouflage, a vampire's lack of reflection is a staple of Western folklore, and it is problematic from a physics standpoint. All material objects should have a reflection, because all matter interacts in some way with light. Different materials absorb, reflect, scatter, or even bend light to varying degrees, depending on their density. For instance, light travels faster through air than it does through water or glass. As light passes from a fast medium to a slower one, the light ray bends toward the boundary between the two, and how much it bends depends on a property called the index of refraction. The index of refraction for water is 4/3, implying that light travels three quarters as fast in water as it does in a vacuum. When a pencil is submerged in a glass of water, the pencil appears to bend. This optical effect occurs because water is denser, so light moves more slowly through water than through air. That's also why submerged objects appear to be shallower than they actually are: The light reflecting off them changes its angle at the surface and bends down toward the water.

A similar effect occurs with glass. As light enters a piece of glass at a given angle, one part of the light wave will reach the glass before another, and it will start slowing down first. This causes it to bend in one particular direction, depending on the angle at which it enters. This is the principle behind a lens—an effect known as refraction. Light mostly passes through a clear, flat pane of glass, like a window—this is called transmission—but a tiny number of photons hitting the glass will be reflected, producing the phenomenon of partial reflection. Giles experiences this effect when he encounters Angel in the high school library in "Out of Mind, Out of Sight" (B-1): Giles sees a partial reflection of himself in the glass door of one of the bookcases, but Angel doesn't show up at all. A mirror is basically a piece of glass that has been "silvered." A coating of tin or mercury on one side ensures that instead of passing through the glass, most of the incoming light is reflected; only a tiny amount will be absorbed. That's why Buffy sees her reflection clearly in her bedroom mirror.

For Angel to cast no reflection, either he would have to absorb light to such an extent that none of it is reflected onto the mirror—in which case, he wouldn't be visible to Buffy or anyone else—or all of the light reflected off his body would have to be absorbed by the mirror's glass. But why should this only apply to him, and not to Buffy? (This unique ability would also have to extend to his clothing.) Furthermore, while vampires cast no reflection, they mysteriously show up on camera. There is no good explanation for this. Although light reflects off a mirror and bends as it passes through a camera's lens, both mirror and lens are based on the same underlying physics principles of absorption, reflection, refraction, and transmission—all different kinds of interactions with light. So if a vampire shows up on camera, he or she must also have a reflection.

The writers of the two series seem to be well aware of this contradiction; it becomes a running gag of sorts. Andrew comments on the anomaly in "Storyteller" (B-7) as he is documenting the Scoobies on his camcorder. Spike casts no reflection, and yet he shows up perfectly on film, even agreeing to do a retake in more flattering lighting. This is probably wise on his part. As Cordelia succinctly puts it in "Are You Now or Have You Ever Been" (A-2), upon seeing an old photograph of Angel in the 1950s: "It's not that vampires don't photograph; they just don't photograph well."

SILENT SCREAMS

Vampires also have greater sensitivity to sound. When Angel, Wesley, and Gunn are imprisoned in a dungeon in the alternate dimension of Pylea, Angel is able to hear two guards speaking in low voices all the way down the hall, but Wesley and Gunn can't hear anything ("Over the Rainbow," A-2). That's because Angel is gifted with an ultrakeen sense of hearing. The same is true of crickets. They have tiny hairs on their bodies that can detect the slightest disturbances in airflow—say, from the beating of another insect's wings—as a natural defense against predators like spiders or wasps. Shifts in airflow cause the hairs to rotate in their follicles, and this in turn fires off a neuron. Crickets can detect low-level sound in any direction, which suggests that vampires might possess some sort of similar mechanism to accentuate hearing. Physicists at the Netherlands-based University of Twente are now building artificial hairs as small as one millimeter long—basing their design on cricket hairs—which could one day be used as sensors in hearing aids.

Other creatures in the Buffyverse react to sound in far more interesting ways. In the Emmy-nominated "Hush" (B-4), the Gentlemen are fairy-tale monsters that steal everyone's voices, so their victims can't scream when the creatures arrive to surgically remove their still-beating hearts. This is hardly gentlemanly behavior, especially given the lack of anesthesia. For two nights, Sunnydale resembles deep space: No one can hear the cries of agony. In fact, people can't make any sounds at all.

The induced epidemic of laryngitis is necessary because sound is the Gentlemen's only weakness, and proves to be the key to their destruction. They are extremely sensitive to any kind of noise. Eventually, Buffy figures out how to get her voice back and emits a single loud, prolonged, high-pitched scream that causes the monsters' heads to explode. With a similar effect, in "The House Always Wins" (A-4), Lorne escapes his pursuers in a Las Vegas casino by belting out a long, loud, high-pitched note that shatters all the lightbulbs in the neon signs that adorn the casino's atrium.

Sound can affect material objects—like the heads of creepily cadaverous demons, or the neon lights of Vegas—because it is comprised of mechanical energy. That's also why it requires a medium, like air, in order to travel. Sound waves are pressure waves: the result of a vibrating object that creates a disturbance in the surrounding air. For instance, when Buffy's telephone rings, the ringer vibrates very quickly, sending energy radiating outward through the air. These vibrations disturb the molecules that make up the air. The air molecules push closer together as the object moves one way—an effect known as compression—and then create a space between themselves and the vibrating object as it moves the other way, which is called rarefaction. The motion disturbs the neighboring molecules in turn, creating an outward ripple effect, much like a stone cast in a quiet pond will cause waves to ripple outward from the spot where the stone hit. So sound waves travel in repeating patterns of compressions and rarefactions.

All sound waves have wavelength and frequency, just like light waves—not to mention a little thing called amplitude. The distance between compressions determines the wavelength. Objects that vibrate very quickly create short wavelengths because there is very little space between the compressions, creating a high-pitched sound. Objects that vibrate very slowly create long wavelengths because the compressions are spaced farther apart.

This creates a low-pitched sound. Frequency measures how many crests, or compressions, occur within one second; the measurement of this speed of vibration is called a Hertz (Hz), and 1 Hz is equivalent to one vibration per second. When Giles plucks a string on his guitar while performing in the Espresso Pump coffee shop ("Where the Wild Things Are," B-4), that string might vibrate five hundred times per second, causing surrounding air particles to vibrate at the same frequency, so the sound wave's frequency would be 500 Hertz. Pitch simply denotes those frequencies within the range of human hearing (from about 20 Hz to 20,000 Hz). The faster the rate of vibration, the higher the pitch; the slower the rate of vibration, the lower the pitch.

A sound wave's amplitude, or range of movement, determines its volume. If Giles plucks his guitar string very hard, it still vibrates at the same frequency as when he plucks it softly, but the amplitude is much greater. The string requires more room to move back and forth, therefore exerting much more mechanical energy and producing a louder sound. We measure the energy level, or amplitude, of a sound wave in decibels (db). A human whisper is about 20 db, while normal conversation is around 60 db. The amplified music at a rock concert can reach as much as 110 db. At 130 db, sound will cause pain, and possibly damage the ear. The human eardrum will actually burst at 160 db.

It is not a simple feat for Lorne to shatter the neon lightbulbs, or for Buffy to cause the heads of the Gentlemen to explode. The secret is a precisely tuned frequency, combined with long duration, and lots of decibels. Every material object has a natural resonant frequency at which it vibrates. That's why running your damp finger along the rim of a crystal wineglass produces a faint hum. The shattering effect is called forced oscillation resonance. If an object has a particular natural rate of vibration, and if one pumps in more energy of the same resonance, the object will vibrate so strongly that it can shatter. What's happening here is the result of sound-wave interference. Two sound waves of identical pitch, or frequency, have peaks and valleys that line up perfectly, so their amplitudes add together, doubling the volume of the sound. This in turn causes the glass to vibrate more from the extra mechanical energy. Allow that energy to build up over several seconds, and the glass will shatter.

Complicating matters is the fact that not all glass is created equal. Different kinds of glass will have different resonant frequencies, depending on its shape and chemical content. The glass used in windowpanes, for instance, vibrates poorly and has no strong natural frequencies. With crystal, lead oxide is often added to the glass to make it stronger. This gives it a natural resonant frequency, and sometimes that frequency can be matched by a human singing voice, although the actual note required depends on the glass. Crystal wineglasses also have a thin, delicate structure and bell shape that make them especially vulnerable to the propagation of resonant vibrations. The vibrations build up inside the "bell" until they become so powerful that the glass shatters. The volume of the sound required varies from glass to glass as well. Memtek, the company that makes Memorex recording tape, used drinking glasses with a high lead content in their popular commercials showing singers like Ella Fitzgerald breaking glasses with their voices. Those voices were amplified to about 94 decibels, on a par with a jackhammer. Other glasses might require as much as an earsplitting 135 decibels.

So the note Lorne sings must be loud (at least 90 decibels), and prolonged (at least two to three seconds) to allow enough vibrating energy to build up to cause the bulbs to shatter. It must also resonate perfectly with the natural frequency of the glass neon bulbs. If Lorne's note does not match the resonant frequency of the bulbs, he can sing as long or as loud as he likes, with no effect: Most of the sound energy will simply be reflected, and the glass won't break. The same criteria apply to the Gentlemen's exploding heads. Those heads would have to have a resonant frequency perfectly matched to the pitch of Buffy's scream (actually more akin to a very prolonged holler) in order for this to happen. Assuming the heads of the Gentlemen are harder to crack than the neon bulbs, Buffy's sustained scream would probably have to be at least 135 decibels in order to generate sufficient mechanical vibration to cause them to explode.

Rather than going to the trouble of stealing everyone's voices to protect themselves from unexpected screams, the Gentlemen could have employed some form of noise-cancellation technology. Unlike earplugs and sound dampeners, noise cancellation tries to block the unwanted sound at its source, rather than merely trying to prevent it from entering our ears. Once again, it's a matter of interference between sound waves, although in this case, the opposite of perfect alignment occurs. If we add two waves together, and the peaks of one line up with the valleys of the other, they will cancel each other out. Digital signal processors (DSPs) are microelectronic devices that determine which sound wave is required to cancel the unwanted sound wave (noise). It then creates that sound and amplifies it through speakers or headphones. The end result is near silence. Most cell phones, CD players, and hearing aids now contain one or more DSP devices.

The She-Mantis who targets Xander in "Teacher's Pet" (B-1) isn't quite as sensitive to the mechanical vibrations of sound as the Gentlemen, but she does exhibit extreme sensitivity to high-pitched sounds—those with higher frequency, as opposed to greater amplitude. Buffy defeats her by playing a recording of bat sonar, which Giles likens to being "soothingly akin to having one's teeth drilled." This supposedly wreaks havoc with the

creature's central nervous system, making the mantis more vulnerable to attack.

Ultrasonic bat sonar does seem to affect the real insect's central nervous system—just not in the way it affects the She-Mantis. Praying mantises have been observed to go into spiraling dives or tailspins when flying in close proximity to bats, which seems to indicate that they somehow lose their bearings in response to the ultrasound. But the aerial dive is actually a defense mechanism against its primary predator, the bat. (Fighter-jet pilots use similar maneuvers to evade enemy fire.) Bats emit a series of ultrasonic pulses that bounce off objects in their environment. How long it takes for the sound to be reflected back to the bat indicates how close (or far) a given object might be, enabling the bat to orient itself as it flies, and to detect food. Modern sonar technology is based on the same principle. The more feedback the bat receives, in terms of incoming reflections, the more accurately it can pinpoint a given object's location. That's why the rate of the ultrasonic calls increases as the bat nears its prey, climaxing into a "feeding buzz" as the bat locks in on its target and prepares to strike.

The praying mantis has a single ear—located in the center of its chest—that is highly sensitized to ultrasonic vibrations, better enabling it to evade a hungry bat. David Yager, a neuroscientist at the University of Maryland, College Park, implanted electrodes into the head of a praying mantis and placed it in a dimly lit "flying room," releasing a bat and recording the attacks on high-speed video and ultrasound detectors. He found that when a mantis picks up the bat sonar, the ear sends a signal to the brain along the auditory nerve. The insect is able to measure the rate of the ultrasonic pulses as it increases in order to calculate the right moment to dive.

NATURAL SELECTION

The She-Mantis kills her victims by biting off their heads which, if nothing else, makes death quick. Certain other monsters aren't quite so merciful. In "Same Time, Same Place" (B-6), we meet Gnarl, a skin-eating demon who first paralyzes his victims by scratching them with long talons that secrete a powerful neurotoxin. This gives Gnarl plenty of leisure time to skin his victims alive slowly and lap up any blood as his "natural beverage." He scratches Buffy's sister, Dawn, first, paralyzing her. But Buffy and Xander manage to rescue her from the creature's cave before it begins to feed. Willow is less fortunate: She is trapped in the cave with Gnarl for a good while before she is rescued, lying paralyzed (and fully conscious) as the thing begins to strip off small bits of skin and eat them.

Neurotoxins are chemicals that change the way the neurons in the brain function, either by inhibiting the release of certain neurotransmitters, or enhancing them to harmful levels. The human central nervous system is a very complex, highly integrated information-processing and control network. The brain is made of 100 billion nerve cells called neurons, which gather and transmit electrochemical signals, much like the gates and wires in a computer. Motor neurons control muscle contractions by carrying signals from the central nervous system to the outer parts of the body. Sensory neurons carry signals from the outer parts of the body to the central nervous system. Receptors sense the environment—chemicals, light, sound, and touch—and encode this information into electrochemical messages that are transmitted by the sensory neurons. And the interneurons connect the various neuron "family members" with the brain and spinal cord.

Neurotoxins wreak havoc with those critical electrical and chemical signals to and from the brain, interrupting speech, thought processes, motor function, even respiration.

There are many different creatures in nature that make use of powerful neurotoxins capable of paralyzing and killing their victims—ranging from the Japanese pufferfish and blue-ringed octopus to the Australian paralysis tick and several species of snake and spider. Maculotoxin is a venom produced by the salivary glands of the blue-ringed octopus. Its prey is often fully conscious and paralyzed as the octopus consumes it, much like Willow was conscious, yet helpless, when paralyzed by Gnarl. Yet both Willow and Dawn can still breathe, and even choke out the occasional sentence despite not being able to move their lips. Technically, the paralysis should spread to the respiratory system. Victims of maculotoxin poisoning often survive only if they receive oxygen immediately; otherwise they succumb to respiratory failure. And there is no known antidote in nature: Killing the offending creature won't magically reverse the paralysis, as Buffy succeeds in doing with Gnarl by poking out its eyes.

Maculotoxin is the venomous form of tetrodoxin, a neurotoxin found in the gonads, liver, intestines, and skin of the pufferfish. In Japan, pufferfish (known as *fugu*) are eaten as delicacies for the tingling sensation the toxin produces on the lips and tongue. *Fugu* chefs must be specially trained to prepare the fish so that only small, nonlethal amounts of the poison remain. Nonetheless, between 100 and 200 people are poisoned each year from consuming pufferfish, and about half of those die. Again, the victim may be fully conscious shortly before death, even though completely paralyzed.

Poison from the pufferfish also plays a role in Haitian zombification rituals. According to legend, zombies are essentially artificially reanimated corpses, but unlike vampires, they continue to decay, and are far from being truly sentient. They are sprinkled throughout the Buffyverse. Buffy's homecoming party is overrun by zombies in "Dead Man's Party" (B-3), and Xander finds himself part of a gang of zombified bad boys in "The Zeppo" (B-3). When the employees of Wolfram & Hart are massacred in "Habeas Corpses" (A-4), many are reanimated as zombies. And zombie policemen terrorize crime-ridden communities in "The Thin Dead Line" (A-2).

In the real world, the secret to creating "zombies" lies in the power of suggestion, combined with such poisonous toxins as those produced by the sea toad (*bufo Marinas*) as a defense mechanism. Local legend holds that the *boko* (voodoo priest) poisons a victim and buries him in an aboveground tomb, in accordance with Haitian custom. The *boko* then steals the body, and revives, then enslaves that person. Anyone who's read *The Serpent and the Rainbow* by Wade Davis (as opposed to viewing the cheesy horror flick by the same name) should recall that the toad is placed in a jar with a stinging sea worm. The two creatures fight, causing the toad to produce plenty of the toxin. This is mixed with a dab of tetrodoxin from the pufferfish. The victim becomes sufficiently catatonic to be mistaken for dead, and is buried. The drug wears off after a few days, and the "zombie" is exhumed and put to work. Regular doses of *datura stramonium*, an extract of the thorn apple, ensures that a "zombie" remains passive and dutifully follows orders, whether those orders are to labor in the fields without complaint, or to eat the brains of the living.

These practices (apart from the brain consumption) were confirmed by British researchers in the late 1990s, although they didn't find any examples of real zombies. More often than not, suspected "zombies" are actually individuals with diseases that cause them to behave erratically and frighten superstitious neighbors, who then denounce them as cursed creatures. One suspected case was a woman who suffered from catatonic schizophrenia, while two others suffered from a severe learning disorder and an organic brain disorder, respectively. But the physiological effects of the various poisons used in the Haitian ritual are quite real, and well documented.

In late June 2005, tabloid headlines in England and Australia screamed about the creation of "zombie dogs" by a group of Pittsburgh scientists. (It's probably just coincidence that the classic zombie horror film *Night of the Living Dead* was filmed in the city's Monroeville Mall.) As with most tabloid stories, the moniker was an overhyped misnomer, but the results described were surprisingly accurate. Scientists have made substantial progress in achieving something called suspended animation with delayed resuscitation. It's not zombification per se, but it is tantamount to reviving the recently dead. Suspended animation is an outgrowth of the field of cryogenics and its cousin, cryonics—the storage of human bodies at extremely low temperatures in hopes of one day reviving them. It's similar in principle to how people who fall into icy water can survive for up to an hour despite the cold temperatures. The body goes into suspended animation: the metabolism and brain function slow down to the point where little to no oxygen is needed.

Researchers at Pittsburgh's Safar Center for Resuscitation Research have developed a new technique for suspended animation that could help save lives of accident victims or soldiers on the battlefield—or anyone who has suffered a lethal hemorrhage.*

*Roughly 50,000 Americans die every year from hemorrhage, which is also the leading cause of death among troops killed in action.

They tested their technique on dogs. First, they drained the animals' blood and replaced it with an ice-cold salt solution, mixed with small amounts of glucose and dissolved oxygen. The dogs were clinically dead, with no heartbeat, respiratory functions, or brain activity, but their tissues and organs were perfectly preserved because the procedure lowers the body temperature to about 50°F. After three hours, fresh, warm blood was pumped in and the animals were revived with electric shocks. Two thirds of the "zombie dogs" in the study suffered no brain damage from the procedure.

Cryogenics pops up in the Buffyverse as well, although it's applied a bit differently from how it was in Pittsburgh. Warren, Jonathan, and Andrew are three former high school nerds who decide to become "crime lords" and take over Sunnydale. For all their geeky ineptitude, they do come up with some creative inventions, including a "freeze ray" device. In "Smashed" (B-6), the Trio (as they call themselves) steals a diamond from the Sunnydale museum, only to be confronted by a humble museum guard named Rusty. They hit Rusty with a blast from the prototype freeze ray, turning him into a block of ice in a state of suspended animation.

In reality, three hours is the upper limit for how long a body can remain in suspended animation and still return to normal function. That's usually long enough to transport trauma victims to a hospital and enable surgeons to locate and repair internal bleeding. Poor Rusty remains cryogenically frozen for several days before he unthaws with the help of a hair dryer. Since Rusty suffers no ill effects from the freeze ray, we can only assume that the science of cryogenics in the Buffyverse is significantly more advanced than our own. For instance, the Trio must have figured out how to keep the cells of living tissue from shattering once they've been frozen. This is one of the biggest challenges facing the field, and a big part of why, to date, no one in long-term cryonic suspension has been successfully revived.

That's because of the physical changes that take place when something freezes. It's a well-established scientific fact that the same substance will behave differently at various temperatures and pressures. Water (H_2O) is the most familiar example. It can be a solid (ice), a liquid (water), or a gas (steam), but it is still made up of molecules of H_2O, so its chemical composition remains unchanged. At sea level, water freezes at 32°F (0°C) and boils at 212°F (100°C), but this behavior changes at different altitudes because the atmospheric pressure changes. In fact, get the pressure low enough and water will boil at room temperature. The critical temperature/pressure point at which H_2O changes from one form to another is called a phase transition.

Cells contain a lot of water. Water expands when it freezes into a crystalline solid, and this expansion causes the cell walls to burst or shatter. So in order to cryogenically freeze a body with any hope of resuscitation, one must first remove all the water from the cells and replace it with the equivalent of human antifreeze, like the Pittsburgh researchers did with their canine subjects. This keeps the organ and tissue cells from forming ice crystals at extremely low temperatures and puts the body into suspended animation so that it can then be cooled down to freezing levels gradually. That's not the only difficulty. The warming-up process must be done gradually at very precise speeds—again, to prevent cells from shattering. A hair dryer might be a nice comic touch, but it wouldn't offer sufficient control over the process in the real world.

Cold isn't the only way to freeze someone in his tracks. When it comes to suspending animation—or at the very least, limiting motion—don't underestimate the power of mucus. When the Chaos-worshiping Ethan Rayne turns Giles into a Fyarl demon ("A New Man," B-4), Giles gains the ability to shoot a sticky mucus through his nostrils that hardens into a solid and immobilizes an opponent—shades of the infamous Queller demon in "Listening to Fear" (B-5). Giles was hoping for a more exciting ability—say, shooting searing laser beams from his eyes. Excreting large globs of snot just doesn't have the same cachet. Giles's disappointment might be assuaged by the news that mucus has some very desirable properties of great interest to scientists—not to mention to celebrities of advancing years. Slug mucus, for instance, is enjoying a renaissance of sorts as an anti-aging compound in high-end cosmetics. Beyond the pursuit of vanity, it also provides a useful model in the development of new synthetic lubricants, which could one day be used to combat friction in molecular-scale nanomachines.

Queller and Fyarl demons might be figments of Whedon's prodigious imagination, but there is solid science behind the slime to explain, for example, why it hardens. It's what's known as a "phase-change" material because it moves from liquid to solid. The change is usually triggered by temperature (hot to cold, or vice versa) or environmental factors (wet to dry, dry to wet). Mucus is made up of protein-and-sugar molecules (mucin), as well as lots of water, which gives the material its slippery texture. As the substance loses moisture, it becomes more rigid, undergoing a sort of phase transition, although scientists who study these strange materials prefer to describe the process in vaguer terms: The substance goes from a "fluid-like" state to a "solid-like" state. When the Queller demon sprays its mucus onto the face of Buffy's mother, Joyce, the substance rapidly cools down and begins losing moisture. As it dries out, it forms a hard shell over her mouth and nose, impeding her ability to breathe. Had Buffy not intervened in time, Joyce would have suffocated.

Earlier, we met the lowly hagfish, whose slime is of great interest to a team of scientists in Boston and British Columbia because of its unusual properties. Unlike other forms of mucus, hagfish slime doesn't harden. It stays slimy even in very chilly water, in part because both the hagfish and its victim are immersed in salty seawater, so the slime never has a chance to dry out. Hagfish slime also has a secret ingredient. The usual protein-and-sugar concoction contains long threadlike fibers. These fibers form protein strands that expand rapidly once the mucins come into contact with seawater, causing the substance to "blow up" into a sticky gel. The consistency is a bit like half-solidified Jell-O or watered-down hair gel. The resulting slime bonds to the gills of an attacking fish and blocks respiratory flow: the victim perishes by choking on snot, just like the targets of the Queller demon. It might kill fish, but this unique aspect of hagfish slime could help save human lives by curtailing bleeding in an accident victim during surgery. The mucus would expand upon contact with the blood (which is mostly water and salt), staunching blood flow. More frivolously, a group of students in British Columbia figured out how to use hagfish slime as an egg substitute in scones; they believe it could also serve as a thickening agent in eggnog.

And of course, as any hagfish could attest, mucus remains a terrific defense mechanism, which is one reason that the U.S. military is investigating its properties for military applications, such as clogging up the blades of an enemy ship's propellers. Who can fight effectively when encased in a foul-smelling, hardened shell of snot? The hagfish gets out of its own mess by tying itself into a knot, then pushing the knot down the length of its body to scrape off the slime. Unless a Fyarl demon's target is equally flexible, said target won't be able to employ a similar means of escape. At least it will have age-defying skin as it succumbs to inevitable defeat.

2

CONDUCTIVITY UNBECOMING

The Shocking Truth About Electromagnetism

"I've just been ionized, but I'm okay now."
—BUCKAROO BANZAI

"Ground State" (A-4) opens with a little girl named Gwen being dropped off by her parents at a boarding school for children with "special needs." What makes Gwen special isn't readily apparent, but she is swathed in layers of clothing topped by a bright red parka with heavy mittens, despite the warm day. Both her parents and her new teacher refuse to touch her, and she is kept isolated from the other children. One day during recess, a little boy comes up and decides to be her friend, holding out his metal toy car as a gesture of goodwill. Gwen removes her mitten to take it—and the little boy is electrocuted, his toy car lying blackened and melted in the grass as the horrified teacher comes running, too late, to intervene.

Gwen possesses an innate ability to manipulate electromagnetic fields.* For most of her life she is labeled a "freak," incapable of even casual human contact—at least not without frying the contactee. Yet she is smart enough to exploit her unique ability to make a very lucrative living as a thief for hire, stealing rare mystical artifacts. After all, for someone like Gwen, high-tech electronic security systems are a snap to disable: Surveillance cameras, electrified gates, laser sensors, even handprint recognition scanners all rely on electromagnetism. She's the human equivalent of a so-called e-bomb, which is capable of wiping out the circuitry of all electrical objects within its range by means of a strong electromagnetic pulse. Gwen merely injects a large influx of electrical current into the wiring, overloading the circuitry and knocking out the entire system.

Electricity and magnetism have a perfectly reciprocal relationship; they are flip sides of the same coin. An electrical current is nothing more than a charge that has been set in motion by the force of a jiggling magnetic field. In fact, modern power plants generate electric power by moving giant coils of wire near magnets. Moving charges, in turn, create more magnetic fields. That's why an electrical current can deflect a compass needle. Electromagnetism permeates the Buffyverse as much as it does our own world. After all, it is a key enabler of most modern technology, from basic appliances like microwave ovens and telephones to advanced computing and robotics.

The inhabitants of the Buffyverse repeatedly demonstrate their

*Gwen's last name is Raiden, a geek in-joke. There is a Japanese thunder god by that name, who also inspired a lightning-wielding Mortal Kombat character named Raydon.

innate understanding not only of how electricity works, but how it can best be used—for good or evil. Among other things, it makes for a terrific weapon. In "Revelations" (B-3), the Glove of Mynhegon is a mystical weapon that draws down lightning from the sky and temporarily stores it, just like an electrical capacitor. The wearer can then redirect the charge to zap intended targets into oblivion. The Initiative uses stun guns and tasers to control and capture demons. In "I Robot, You Jane" (B-1), two computer nerds conspire to electrocute Buffy in the girl's locker room. (The shower floor is wet.) A similar incident occurs when Buffy is ambushed in a sewer by a demon ("The 'I' in Team," B-4). Buffy's electrical stun gun malfunctions, but the demon is standing in a puddle of water. She tosses the gun into the water to electrocute it.

OPPOSITES ATTRACT

While Buffy is being ambushed in the sewer, Initiative operatives hit Spike with a tracer dart. He reluctantly turns to the Scoobies for help. Willow casts a magic spell that ionizes the atmosphere around Spike, interfering with the electromagnetic signal from the tracer to throw off the Initiative's soldiers long enough to remove the dart. The suddenly ionized air causes all the lightbulbs in the room to explode—and the Scoobies' hair to stand on end.

Ionization is the secret to electrical conductivity. The basic unit of electricity is charge. There are two kinds of charge, positive and negative. Like charges repel, and opposite charges attract. Atoms are made of positively charged protons and neutrons (which have no charge), bound tightly into a nucleus and orbited by negatively charged electrons. Normal atoms have exactly the same number of electrons and protons, so they are neutral. But sometimes an electron is knocked out of its orbit, and the atom becomes imbalanced. A charged atom is called an ion. It has a positive charge if it loses an electron, or has a negative charge if it gains one. Like most things in physics, charge is conserved, which means that the total electric charge in the universe is zero. So if an atom loses an electron and becomes positively charged, this must be balanced elsewhere so that the sum-total charge equals zero. Electricity is simply a current of electrons flowing from atom to atom, trying to balance things out. How well they manage to do so depends in large part on the particular substances they encounter.

Electrons are more easily transferred from atom to atom in water, for example, or in certain metals like copper, silver, or aluminum. That's why Gwen electrocuted the little boy who wanted to be her friend: The metal toy car served as a conductive conduit for a powerful surge of electric current—a large-scale version of the small static shock one receives after scuffing one's feet on the carpet and then touching a metal doorknob. Charge is built up and then transferred. Metallic materials are known to be good conductors. In contrast, materials like rubber, glass, and plastic are insulators: Charge can be transferred to and from the surface but does not move easily among the atoms throughout the material. When Gwen's double-crossing client outfits a special plastic elevator to trap her, there is nothing metallic to conduct electricity, so Gwen is powerless to escape on her own.

Gwen's ability most likely stems from the same mechanism that causes lightning to form and strike the Earth. We've already seen that there is a link between electromagnetism and light: Whenever a magnet jiggles or an electric current changes direction, a wave of radiation—or light energy—spreads outward into space. In much the same way, a charged particle creates a field around itself that fills the surrounding space, and thus it is capable of exerting a force on other charged particles within that field. Since the strength of a physical force decreases with distance, the closer the charged particles are to each other, the stronger the forces are between them. And sometimes those forces lead to a polarization of charges in a given object, like within a storm cloud.

Storm clouds form as moisture evaporates from the ground into the atmosphere, where the droplets congregate and jostle against one another, like the young patrons at the Bronze on a busy weekend night. As more and more water droplets collide inside a cloud, their atoms bounce off one another more forcefully. This atomic slam-dancing knocks off electrons. The ousted electrons gather at the lower portion of the cloud, giving it a negative charge, while the upper part of the cloud becomes positively charged.

Eventually the growing negative charge becomes so intense that electrons on the Earth's surface are repelled and burrow deeper into the Earth. The Earth's surface becomes positively charged, and hence very attractive to the negative charge that's accumulating in the bottom of the cloud. All that is needed is a conductive path between cloud and Earth, in the form of ionized air.

The human body is essentially a reservoir of positive and negative charges, and it is normally neutrally charged. Gwen's body may have an overall neutral charge, but it is severely polarized, just as clouds are when they give rise to lightning. Not surprisingly, Gwen has been struck by lightning fourteen times. Her negative and positive charges separate into different parts of her body to such an extent that the "lightning" effect occurs all the time, and much more intensely. It's rather like what happens with an electric eel, a creature capable of producing strong electrical currents as high as 650 volts to stun or kill its prey. The eel is essentially a naturally occurring battery. Its tail end has a positive charge, while its head region is negatively charged. So whenever the eel touches its tail and head to other animals, it sends electric shocks through their bodies.

Consider the ill-fated little boy with the toy car. The negative charges in the toy car are pushed farther away, repelled by the negative charge in Gwen's outstretched hand, so that the positive charges are in the foreground. Gwen's negative charge, in turn, will be attracted to the growing positive charge in the toy car. When she makes contact, this creates a conductive path, causing a surge of electrons to flow from her fingers through the toy and into the boy's body. Normally we would experience this as a small static shock. Even though the human body is about 70 percent water, with a great deal of salt—salt water is an excellent conductor—the fats, oils, and proteins that encase the cells serve as barriers, allowing for only a little electrical conductance. But once the current passes into the bloodstream, there is almost no resistance, because blood plasma is made up almost entirely of salt water. The current pouring from Gwen is strong enough to bypass the protective cell membranes entirely, electrocuting the little boy.

A lethal dose of electricity can paralyze the respiratory organs, damage the central nervous system, and cause burns where the current enters and leaves the body. It is also powerful enough to stop a human heart, and this is what most often proves fatal. The heart beats in response to a series of carefully timed electrical impulses. Throw that rhythm off far enough, and the result is cardiac arrest. This is what happens to Gunn when he and Angel try to stop Gwen from escaping with a stolen mystical relic for one of her black-market clients. Fortunately, Gwen remembers the little boy she killed as a child and decides to be merciful, administering a second shock to start Gunn's heart again. ("Just like starting a Chevy," she says with a shrug.) She even accidentally starts Angel's long-dead heart briefly as they exchange blows.

The confrontation occurs because Team Angel is attempting to steal the same mystical item, which is housed in a room-sized

vault, protected by laser trip wires in the security system. The laser beams are—correctly—invisible until Angel sprays an aerosol mist into the room to show where they are. We can't see light unless it reflects off something. For instance, the light from a laser pointer is only visible at the spot where the beam hits the screen. When Angel sprays the aerosol into the vault, he releases tiny particles of matter into the atmosphere. The laser light reflects off those particles so that the beams of light can be seen.

The laser beams prove to be the least of Angel's problems. As he is waiting for Fred to disable the high-tech security system for the next stage of the heist, Gwen appears, lowering herself down from the ceiling and (supposedly) using her ability to manipulate electromagnetic fields to curve the laser beams inward and upward, leaving her space to access the relic without tripping the alarm. Gwen explains to the dumbstruck Angel that she is "exciting the subatomic particles with electrical energy and bouncing them off each other before they hit ground state."

This is an accurate description of how atoms absorb and emit energy, but does nothing to explain how she accomplished the feat. Laser beams are made of photons, and light doesn't bend as a result of electrical excitation; it travels in a straight line. Fiber optic telecommunication systems must use tiny mirrors and specialized waveguides to get light to turn corners. It is certainly possible to deflect charged particles using the electromagnetic force, but photons are not charged particles.

In fact, the only force capable of bending light is very strong gravity—an effect first predicted by Albert Einstein in 1915 and confirmed in 1919 by scientific observations during a solar eclipse. According to Einstein, it's not so much that light is bent, but that space-time curves, and light merely follows that curvature. The most dramatic example of this is gravitational lensing, when the light from a bright, distant star is blocked by a massive object, such as a galaxy. The star's light is bent by the galaxy's gravitational field, and astronomers see twin images of the same star. The intervening galaxy acts as a lens, focusing the image of the star to a new location.

Gwen has power over electromagnetism. Unless she has also developed the ability to generate a strong gravitational field, sufficient to bend space-time at will, she couldn't bend the laser beams. But perhaps she has found a way to create her own version of a waveguide out of ionized air. We saw in the previous chapter how light moves through materials at different speeds, depending on their respective indices of refraction. If Gwen were able to transfer enough electric charge from her highly polarized body into the air surrounding the laser beams, that air could become sufficiently ionized to change its index of refraction, essentially turning it into a form of matter known as a plasma (an ionized cloud of gas).

The catch is that Gwen must control this process precisely enough to create localized "patches" of ionized air that form "paths" in the configuration of how she wishes the laser light to bend. Those patches would be hotter, with a much lower index of refraction than the surrounding cooler air, and these differences would create a "wall" or barrier. The laser light would bend and follow those ionized paths much like lightning wends its way toward the ground or water runs through a rubber hose. Gwen could then make the laser light do whatever she wanted. Researchers at the Rutherford Appleton Laboratory in England have done something quite similar, using tiny ceramic capillaries (tubes) to confine the resulting plasma. They call their structure a gas-filled capillary discharge waveguide.

It's doubtful, however, that Gwen possesses that level of control over her ability. If she did, she would be able to control the polarization of charge in her own body—and she can't, not without some technological assistance. In "Players" (A-4), Gwen is seeking to steal a new technology called Localized Ionic Sensory Activation (LISA),* an implantable device that is designed to regulate the body temperature, chemistry, and heartbeat of covert operatives. This technology is perfectly feasible. There is a similar real-world implant under development that controls blood pressure in the patient. The implant sends electrical signals to the central nervous system, which are interpreted as a rise in blood pressure. The brain then counteracts this perceived increase by dilating blood vessels to allow blood to flow more freely, reducing the heart rate. LISA would operate on similar principles.

For once, Gwen isn't being mercenary. She is simply tired of being a freak. She desires LISA because it can neutralize her own body chemistry so that the deadly polarization of charges balances out. The implant works. With LISA in place, she can touch another person without electrocuting him or her. Thanks to LISA, Gwen finally finds her ground state.

VANISHING ACT

Gwen might be grounded, at least temporarily, but there's plenty of other electromagnetic mischief to be found in the Buffyverse. In "Gone" (B-6), the Trio perfects an invisibility gun, described as a "particle ionization" device. But when Buffy walks by unexpectedly, Jonathan and Andrew panic and accidentally set off the

*Not to be confused with the real-world Laser Interferometer Space Antenna (LISA), designed to hunt for gravitational waves. LISA is slated for launch around 2015.

gun. Buffy is instantly rendered invisible, along with a nearby traffic cone.

Buffy enjoys her unusual state at first, playing pranks and relishing the opportunity to hide in plain sight. But she was hit with a massive, uncontrolled dose of radiation from the invisibility gun, and the radiation is causing her cells to mutate at an accelerated rate. Eventually her molecular makeup will lose its integrity and she will disintegrate into a gooey substance with the consistency of tapioca pudding. Fortunately for Buffy, the effect is ultimately undone by a second blast from the invisibility gun, this time on the reverse setting.

Obviously there is no such thing as an invisibility gun—certainly not one that can reverse the process on cue. Particle ionization is a genuine physical effect, but it does not render objects invisible. However, there is such a thing as ionizing radiation that can irreparably damage living cells. We've already seen how the relatively low energy of the sun's UV radiation can damage and kill skin cells. Both X-rays and gamma rays are forms of light with extremely short wavelengths and high energy. At such frequencies, light is powerful enough to knock electrons off of atoms, giving them a positive charge instead of a neutral charge and thereby creating ions. The resulting free electrons then collide with other atoms to create even more ions. An ion's electrical charge can lead to unnatural chemical reactions inside living cells. It can break DNA chains, causing the cell to either die or develop a mutation and become cancerous—not exactly tapioca pudding, but often fatal nonetheless.

The early X-ray pioneers, unaware of the dangers of excessive exposure, suffered burns, numbness, infection, diseased lymph nodes, and cancerous tumors. Gamma rays, released by nuclear reactions, are even more powerful. Whenever the atomic nucleus of a radioactive element like uranium, for example, "decays"—

changes into another element—it emits a small amount of excess energy. Usually the energy is released gradually, but a nuclear bomb sets off a chain reaction, releasing huge amounts of gamma radiation all at once. Depending on the intensity and duration of exposure, the effect on living cells can be devastating. Studies of survivors from the Hiroshima and Nagasaki bombings revealed such symptoms as nausea and severe vomiting; cataracts in the eyes; severe hair loss; and loss of blood cells. Many later developed leukemia or other cancers.

An invisibility gun might not be feasible, but scientists do have some ideas about what it would take to render a solid object like Buffy undetectable to the human eye. Most manmade "cloaking" schemes rely on the camouflage principle. These usually involve light sensors that create a mirror image of the background scene on the concealed object. For instance, a scientist named Susumu Tachi invented his own "invisibility cloak" that projects a computer-generated image onto itself of the scene directly behind the wearer. The viewer effectively sees "through" the cloak, even though it is completely opaque. Researchers in Tokyo are developing a similar fabric in which the background is projected onto light-reflecting beads embedded in the material. And in conjunction with the U.S. Army, scientists at the New Jersey Institute of Technology are developing a special smart coating embedded with tiny microscopic electromechanical machines to enable tanks to change color on the battlefield, making themselves virtually invisible by blending perfectly with their environment.

But mere camouflage has its limitations. It is highly dependent on viewing angle, for example. One of the latest and most intriguing concepts for rendering objects invisible involves electrical waves that travel along the surface of some metals. When light strikes a metal like silver or gold, it generates electron waves, called plasmons. In 2005, researchers at the University of Pennsylvania figured out how to use plasmon coatings as a cloaking device to render solid objects invisible—or nearly so—to an observer. We see objects based on the visible light that is scattered off their surfaces, so if we can prevent light from bouncing off the surface of an object, the object would seem to disappear. A plasmon coating essentially comprises a "shell" around an object. It limits light scattering, because the plasmons resonate at the same frequency as the light striking them. The two are literally on the same wavelength. When this happens, the two frequencies cancel each other out and the plasmon-coated object scatters very little light, making it extremely difficult to detect. The technology would serve roughly the same purpose as the mystical tattoos that renegade lawyer Lindsey McDonald inks onto his entire body in Season 5 of *Angel*. Those tattoos enable him to escape detection by Wolfram & Hart's powerful senior partners, and also to evade the law firm's high-tech security systems.

Plasmons sound like an ideal solution. Unfortunately, the effect only works with wavelengths of light roughly the same size as the object one wants to render "invisible." If an object is bigger than the wavelength of light that shines on it, that object will necessarily be visible, because it will disturb the light, causing it to scatter. So only microscopic objects would be invisible to human eyes. Pint-sized though she is, Buffy is still many times larger than the range of wavelengths for visible light—the only portion of the electromagnetic spectrum our eyes can detect. Ergo, she would still be visible. Buffy is also not metallic, and the effect to date has only been observed in metals. However, a plasmon coating could render Buffy "invisible" to radio waves. She would then be undetectable to radar systems, for example.

The shape of the object is also problematic. Thus far, scientists have only managed to achieve the effect with spherical or cylindrical metal objects in the nanoscale range. Plasmon light-shielding covers would have to be customized to match the properties of every object they hide, and be able to cope with all wavelengths of the visible spectrum. Currently, the effect only works with a precisely tuned single color of light. Even a small object might not be able to "disappear" in daylight, since natural daylight contains many different wavelengths mixed together. Buffy's shape is hardly perfectly spherical, and bears no resemblance to the size and shape of the traffic cone that was also made invisible. Each would need a separate plasmon coating carefully tailored to each specific wavelength in order to be rendered invisible.

NEWTON'S CURSE

Not only does electrical conductivity allow charge to flow through and along the surface of things, it also holds material objects together. Atomic charge is what makes matter "corporeal." In "Chosen" (B-7), a reformed Spike literally goes out in a blaze of glory by means of a mystical amulet worn around his neck. He becomes a conduit for a white-hot energy that incinerates an army of *über*-vamps, along with Spike himself. But death is rarely the end in the Buffyverse, especially for a much-beloved character with cool hair. His "essence" is trapped inside the amulet, even though his physical body has been destroyed.

Spike reappears on *Angel* in "Just Rewards" (A-5), when the amulet mysteriously arrives at Angel's new offices at Wolfram & Hart and Spike's essence is released. But he still lacks a physical body. He is visible, but incorporeal. This should mean he is a ghost—except he isn't. The question of what, exactly, Spike has become is never answered, although Team Angel's resident physicist, Winifred ("Fred") Burkle, talks about his lack of particle cohesion and muses, "It's almost as if your essence is straddling a dimensional void."

What does it mean to be incorporeal? Etymologically speaking, the term derives from the Latin word *incorporeus,* and describes something that has no material body or form. The problem with describing Ghost-Spike (for lack of a better term) this way is that no truly incorporeal being could interact with its physical surroundings the way Ghost-Spike does. True, he can't touch anyone, he can walk through walls, and when he first emerges from the amulet, he lunges at Angel in a fury, passes right through Angel's body, and finds himself standing in the middle of Angel's desk. Yet he somehow manages to make contact with the floor, and he later makes himself quite comfortable seated in Angel's desk chair—indications of a corporeal being. The willing suspension of disbelief notwithstanding, this is a troubling inconsistency.

Of course, matter itself is not the solid, tangible thing we believe it to be. On an atomic scale, there's a lot of empty space. An atomic nucleus is 10,000 times smaller than the region in which electrons orbit around it. Electrons are restless particles with serious intimacy issues; they need a lot of personal space. The audience members at Caritas, Lorne's demon karaoke bar, behave a bit like electrons. Each person (or demon) occupies only one seat at a given table, and would most likely shove off a stranger who tried to sit in his or her lap. Although, if that person happened to be an attractive member of the opposite sex, the two could share a table.

The tables, each with a certain number of seats, can represent the atomic orbitals we learned about in chapter 1, which correspond to specific energy states. Only a certain number of electrons can occupy a given orbital. For instance, a helium atom has two electrons, and both can occupy the same orbital (or "table"), provided they are opposite "genders" (spin up, spin down). In this case, two is company, and three is definitely a crowd: If a third electron tries to horn in, it will be bumped up to the next energy level. In Caritas, audience members are scattered randomly throughout the room at separate tables, but electrons are more orderly. As an atom absorbs energy and new electrons are added, the electrons fill up the "seats" one by one, beginning with the most desirable "table"—the ground state—and working outward. This is the Pauli exclusion principle: Once an electron occupies a seat, it excludes others from sitting in the same seat. (There is no lap dancing in electron land.) As electrons fill up the shells, their number determines an atom's chemical properties, thereby creating the periodic table of elements.

The attractive forces between oppositely charged particles are what hold matter together, while the repellant forces generated between atoms with like charges are what differentiates one solid object from another. Angel's undead body is made up of billions upon billions of atoms, but he only appears to be "solid" as we think of the term. The same is true of his spiffy executive office chair. When Angel takes a seat, we might assume that the atoms in the chair are colliding head-on with the atoms in Angel's backside. But this is not, in fact, what is happening. Instead, the charged particles in Angel's derriere exert an electromagnetic force field and repel the charged particles in the chair seat, and this force is strong enough to keep him from falling straight through the chair to the center of the Earth. There is no other significant difference, physics-wise, between the atoms of the desk and the atoms of the chair (or the atoms of Angel's behind, for that matter), so why should Ghost-Spike go through one and not the other?

As forces go, electromagnetism is much stronger than gravity, even though it is gravity that holds the planets in their orbits. We see a dramatic example of this in "Shells" (A-5), when Team Angel confronts the ancient demon Illyria. The fight is short-lived: She tosses Angel through the high-rise window of Wolfram & Hart's L.A. headquarters. Gravity causes Angel to accelerate rapidly as he falls, but he then slams into the concrete pavement with a resounding thud. Even after building up all that momentum during his fall, electromagnetism still easily trumps gravity once Angel makes contact with the pavement. The repulsive electrical force between the atoms in the pavement and the atoms in his body stops him cold. It is electromagnetism that provides the resistance. No particle in Angel's body ever makes contact, on an atomic scale, with any particle in the pavement.

In "Hellbound" (A-5), Ghost-Spike figures out how to affect objects on the corporeal plane by focusing his mental energy. It's a neat trick, but unrealistic: The brain communicates through electromagnetism. If Ghost-Spike's "matter" is not integrated with reality, as Fred had asserted, he should not be able to make use of physical forces like electromagnetism. He also shouldn't see or be seen, since this constitutes an interaction with light (electromagnetic radiation). "In physics, two things that both interact with something else will always be able to interact with each other," Lawrence Krauss writes in *Beyond Star Trek*. "Once you interact electromagnetically, you are a part of our world." Krauss dubs this the "Curse of Newton's Third Law": For every action, there is an equal and opposite reaction.

Newton's curse applies equally to the mother-and-son ghosts haunting Cordelia's chic new apartment in "Rm w/Vu" (A-1). The mother ghost is very controlling. In life she had bricked up her own son, Dennis, in the apartment wall to prevent him from moving out to marry his girlfriend, then she promptly had a heart attack and died. The two have been haunting the place ever since. Mama Ghost is hostile, moving chairs, lifting Cordelia's bed as she sleeps, and shattering objects to drive Cordelia away. Once Cordelia defeats Mama Ghost and releases Dennis's spirit, he becomes a loyal invisible friend. But if Dennis and his mother can move objects, they are interacting electromagnetically with the physical world. In order to see Cordelia at all, they must interact with light, and if they can see her, she must in turn be able to see them—all the time, not just when they wish to be seen.

Not even the First (Evil, that is), the ultimate incorporeal being who first appears in "Amends" (B-3), is immune to the third law's power. As an intangible being, it can't be killed, nor can it interact directly with the physical world. It must manipulate people into doing its bidding. But the First *does* interact with the physical world. It is frequently visible, able to take on the form of anyone who has died. So the First must, by definition, interact with light. And since it can be heard, it must be able to generate mechanical vibrations in the air to produce sound waves. If the First is interacting with physical forces in this way, it cannot be truly incorporeal.

There is one potential loophole to Newton's curse. Ghost-Spike begins to vacillate between the "reality" of Los Angeles and a separate spectral dimension that coexists on the grounds of Wolfram & Hart. When he is trapped in this dimension, Ghost-Spike can still see and hear Fred, but she can't see or hear him. Normally, Newton's curse would dictate that if Ghost-Spike can see Fred, she should also be able to see him; they are linked through electromagnetism. In fact, at one point the invisible Ghost-Spike reaches out to Fred in her lab, and his hand creates an electric spark between them, causing her to sense his presence. It doesn't get more interactive than that.

However, this assumes that the spectral dimension is also a three-dimensional world. If Ghost-Spike inhabits four spatial dimensions instead of three,* his purview would be more expansive than Fred's: He could see her, but she wouldn't be able to see him. (The same goes for Cordelia's ghosts, and for the First.) Light would pass beneath him, rather than scattering off him, rendering him invisible, and he could disappear at will and walk through walls. Fred, trapped in a mere three spatial dimensions, would be unaware of this higher plane of existence, even if the fourth dimension was just inches away. Her perspective would be akin to a two-dimensional person on a flat printed page, or fish swimming just beneath the surface of a pond. The fish would not be aware of the world hovering just above the surface of the water, although they would be able to marvel at the reflected rays of sunlight dancing across the surface, much like Fred can occasionally "sense" Ghost-Spike's presence indirectly, even when she can't see him.

This extra-dimensional spectral world may indeed be part of the Buffyverse, along with others like it. One of the texts Fred consults while researching Ghost-Spike's dilemma is a treatise on fractal geometry in twelve-dimensional space, and she mutters to herself about "interdimensional plasma dynamics" while struggling with equations. Of course, this assumes the existence of extra dimensions—a subject of much contention among real-world physicists, albeit an accepted reality in the Buffyverse. And once Ghost-Spike bounced out of the higher spectral dimension and back into ours, he would become visible, and Newton's curse would once again apply.

*Not counting the dimension of time.

MIND OVER MATTER

So not all of the electromagnetic phenomena that appear in the Buffyverse are equally plausible in our own world. Take the case of a seemingly helpless young woman named Bethany, who is cornered in an isolated alley by two thugs in "Untouched" (A-2). A nearby two-ton Dumpster slides across the alley, crushing the thugs against a brick wall with such force, the cops literally have to scrape them off. The Dumpster didn't move of its own accord. Bethany can move physical objects with her thoughts, a power known as telekinesis.

Telekinesis is an ability that seems like it ought to exist. After all, the human brain uses electrical signals to regulate and control body movement, as well as internal functions such as heart rate and blood pressure. Scientists have long been able to measure the electrical patterns of brain waves, using electroencephalography (EEG), a medical-imaging technique that analyzes electrical activity along the scalp that is generated by the brain. So we know that local current flows are produced when neurons are activated by chemical messengers known as neurotransmitters.

Electromagnetism is a physical force, and as such it can (and does) affect the physical world. Any charged particle can be manipulated with electromagnetism. When it comes to telekinesis, however, the problem is largely one of inverse proportion. Force fields spread out uniformly in all directions from the source, and dissipate rapidly over distance. Even in the Buffyverse, physical forces like gravity and electromagnetism are subject to the same "inverse square laws." Angel may seem to defy gravity, able to leap higher than any human, but this is due more to his supernatural strength. He can exert more force to temporarily overcome gravity, but in the end, gravity still pulls him down to earth. An inverse square law simply means that if something is twice as far from the source of a given force, the strength of that force acting on the object will decrease by four times as much. It is a well-established law of physics that energy is conserved. So if a signal has a certain amount of energy, as that signal spreads out through space, the energy per unit area being carried by the signal must necessarily decrease. The same amount of energy is being divided across a greater area. The signal will be strongest at its source, then weaken rapidly as it spreads outward until it dissipates entirely.

This means that in order for Bethany to lift the scarf lying at Cordelia's feet by using just her mind—part of an exercise in learning to control her ability under Angel's tutelage—she would have to radiate a very strong electromagnetic signal in all directions just to generate enough force to raise the scarf off the ground for a second. Even though all she needs is roughly 0.1 watt of power,* in order to exert that much force directly on the scarf she would need to expend about 10 kilowatts of power in all directions. The good news is that, assuming she has developed enough control to focus her mental signal like a laser beam, she might be able to reduce this to a more feasible 10 watts of power. The bad news is that electromagnetic brain waves are much, much fainter than that, and they must be amplified thousands of times in an EEG before they can be displayed on paper or stored on a computer. Unless she has a similar means of amplification, her brain waves just wouldn't be strong enough to lift the scarf off the floor.

An object as heavy as the two-ton Dumpster would require

*In this context, "power" refers to a specific unit of measurement in physics that denotes how much work can be done in one second. Work is energy that has been harnessed for use—say, to move a two-ton Dumpster. We'll learn more about work in the next chapter.

significantly more energy to move it roughly six feet, particularly since Bethany must do so in about one second. Otherwise her attackers would be able to leap out of the way. Assuming a focused signal, and excluding (for simplicity's sake) any additional resistance from friction as the Dumpster scrapes along the asphalt, Bethany would need to put out roughly 8 kilowatts of power in that one second. This is equivalent to the energy consumed each second by eight American households, or four times the power output of a professional cyclist like Lance Armstrong. Human beings are capable of some extraordinary physical feats when under extreme stress, and Bethany is fighting for her life in that alley. Still, the brain alone doesn't put out anywhere near that kind of energy, no matter how dire the circumstances.

The same concept applies to mental telepathy or extrasensory perception (ESP). Willow uses this ability to great effect to direct the Buffy-less Scoobies while patrolling for vampires in "Bargaining" (B-6), tapping into their heads with instructions and helpful tips. Again, it's understandable, in this age of pervasive wireless communication, to conclude that it must also be possible to communicate nonverbally via invisible "thought waves." We communicate via invisible radio waves all the time.

The problem is that this would necessarily involve electromagnetic force,* and therefore we should be able to detect the signal. No scientist to date has succeeded in detecting any electromagnetic waves associated with ESP (or telekinesis, for that matter),

*One could postulate the existence of a mysterious "fifth force," as yet undetected by scientists, but if such a force exists, according to Krauss, its strength in proportion to gravity would have to be less than 1 part in a trillion in order for it to escape detection by our cutting-edge scientific instruments. Since the Buffyverse is fictional, a mystical "fifth force" responsible for telekinesis and telepathy can't be ruled out entirely, but it would still be subject to the same inverse square laws.

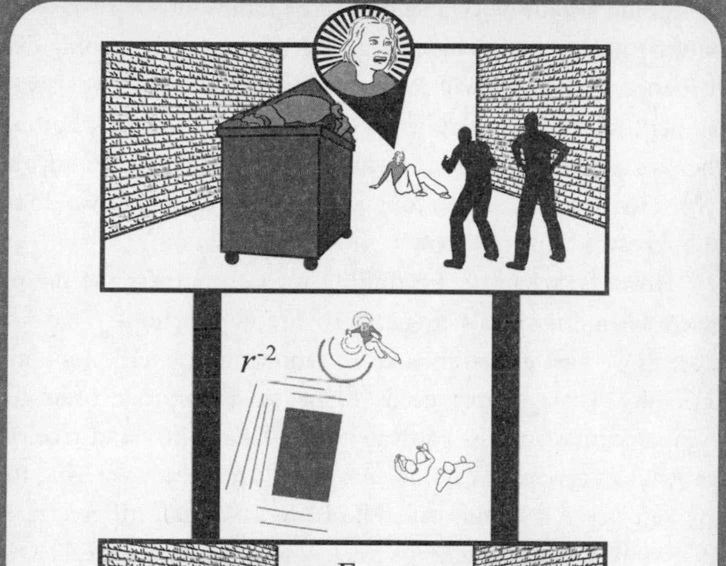

r-2
F = ma
E = 1/2mv2

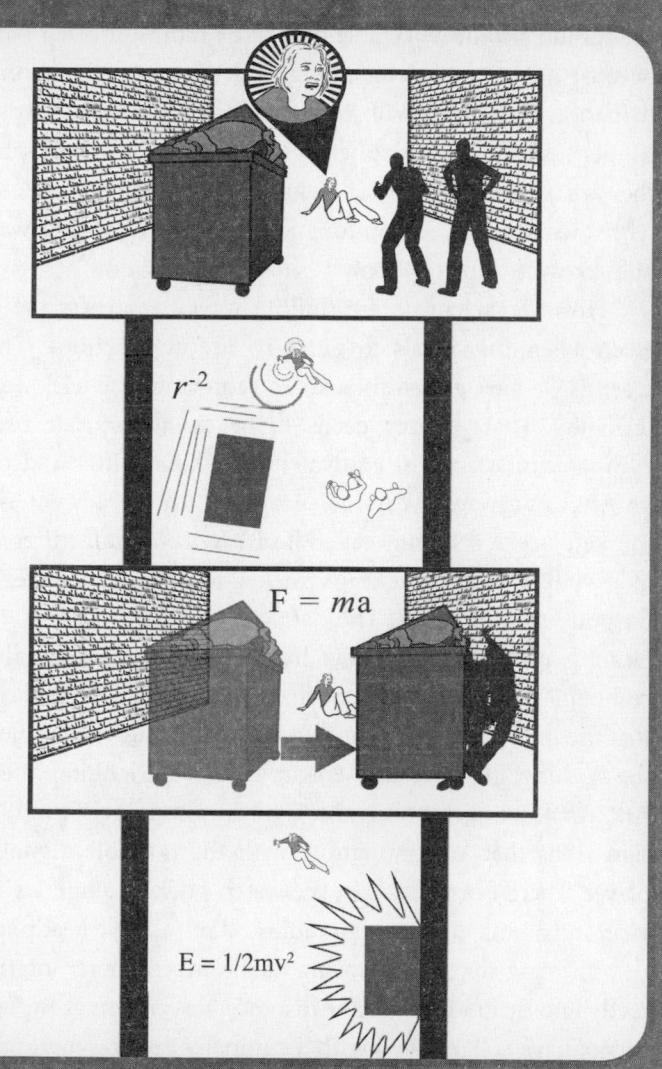

even though they can measure not only very faint brain waves, but also electromagnetic waves at the farthest reaches of the known universe. Any signal being transmitted by Willow's telepathy should also be detectable by nearby radios or other receivers, because it must be strong enough to cause a corresponding disturbance in the brains of Xander and Anya, several yards away. Modern scientific instruments can detect individual photons; they are certainly capable of picking up telepathic signals. Anyone with a wireless connection tuned to the same frequency would be able to eavesdrop on Willow's "silent" conversation.

However, scientists are finding ways to harness the brain's electrochemical signals to perform useful functions. That's as close as we have gotten thus far to achieving true telekinesis and telepathy. There simply needs to be an appropriate brain-to-computer interface, the equivalent of a transmitter and receiver known as wetware. Wearable computers are already entering the marketplace. A company called BodyMedia manufactures a wearable wireless device that keeps track of total calories burned based on body mass changes. The data are uploaded wirelessly to the wearer's computer. Motorola has partnership deals with other companies to develop Bluetooth-enabled sunglasses and clothing that use interactive cell phone and portable music technology; for the clothing, the wires and sensors are woven right into the fabric. And Apple's hugely successful iPod inspired Levi Strauss to design jeans that are compatible with the portable digital music player. There is a joystick in the watch pocket to operate the device, and a built-in docking cradle and retractable headphones.

The next step is implanting the brain–computer interface directly into the body itself. For instance, researchers at Brown University have collaborated with a company called Cyberkinetics to produce an implantable brain-to-computer interface, currently under clinical testing, to enable the user to move a cursor on a computer screen with his or her thoughts, just like one would use a mouse. A sensor is implanted onto the motor cortex of the brain, and the brain signals are then measured and translated into the type of signals useful for computer control, which can be sent to the computer wirelessly. And a research group at Emory University has implanted a transmitting device into the brain of a stroke patient, linking the motor neurons to silicon so that the patient can move a cursor on a computer monitor just by thinking about it.

There are also neurally controlled prosthetics, including a robotic arm that can be controlled using arrays of as many as 96 implanted electrodes that record brain signals. The prosthesis has already been successfully tested in the laboratory, using macaque monkeys. The electrodes were placed in multiple regions of each monkey's cortex, including the area that controls motor function. Scientists at Duke University even managed to transmit brain signals over the Internet in 2000, enabling the monkeys to remotely control the robot arm. If monkeys can do this, why not Bethany? All she would need is the appropriate implant: a very powerful transmitter capable of tuning, amplifying, and focusing the frequency of its signals to move objects of every shape and size. The sticking point is that every single object and person in the Buffyverse must therefore either act as a natural receiver or have a corresponding implant.

Things get even more complicated when it comes to Willow's telepathy. It's not enough for Willow to be powerful enough to transmit a mental signal. She must also be able to receive incoming signals to carry on a "conversation" with her fellow Scoobies, who must be equipped in turn with their own built-in transmitters and receivers. And it still won't solve the problem of eavesdropping by other receiving devices in the vicinity that are sensitive to the same frequencies. Yet the concept is not entirely far-fetched. Electronic sensors are already being embedded in bridges and other structures to monitor damage. There may well come a day when every object and person is "wired" to send and receive electronic signals. Then it just becomes a matter of sending signals with enough electromagnetic energy to move objects. And as we have seen, that energy is considerable.

MAN AND MACHINE

The next logical step in electromagnetic body modification is illustrated in "Lineage" (A-5). Wesley's father, Roger Wyndham-Pryce, shows up at his son's new office at Wolfram & Hart, and the visit coincides with a full-scale attack on Team Angel's new headquarters by a band of cyborg ninja assassins. A cyborg is a cybernetic organism—part human, part machine. When Fred conducts an autopsy of one of the fallen cyborgs, she discovers that their central nervous systems are human and intact, but all their organs have been replaced with advanced cybernetics.

The field of cybernetics is about "humans and technology operating together," according to Kevin Warwick, a professor of cybernetics at Reading University in England and author of *I, Cyborg*. Warwick made global headlines in 1998 when he had a silicon chip transponder surgically implanted into his left arm, turning himself into a living cyborg experiment. The chip communicated wirelessly via radio waves with a network of antennas throughout the building that houses the School of Systems Engineering. These in turn transmitted the signals to a computer in his building. The implant identified Warwick to the computer, which proceeded to open doors and switch on lights whenever he was nearby.

The first implant was in place for nine days. In 2001 Warwick had another implant wirelessly connecting his nervous system (via the nerve fibers in his arm) to the computer. This nerve center carries more information than any other part of the anatomy, apart from the spine. When Warwick moved a finger, for example, an electronic signal traveled from his brain to activate the muscles and tendons that move his hand. The nerve impulses still reached the finger so it could move, but scientists could tap into the signals, recording and storing the data on a computer for later analysis.

Warwick isn't working in a vacuum, even if he's received the most media exposure. Research into electronic implants is booming, in part because of the potential payoffs, both financial and in terms of improved human health. Among other advances, artificial retinas—sometimes called "bionic eyes"—have been successfully implanted in human subjects. The device stems from the discovery several years ago that electrically stimulating the retina could cause some visual perception in blind people. It pairs a tiny electronic eye implant with a miniature video camera that is mounted on a pair of sunglasses. The camera captures visual signals and transmits them to the retinal implant via a wireless receiver embedded behind the ear. The implant connects damaged photoreceptors—rods and cones—on the patient's retina. The incoming signal is then re-created by the electrodes as they stimulate the photoreceptors, which transmit signals to the brain through the optic nerve.

To help patients whose central nervous systems have been damaged by diseases like multiple sclerosis, other research groups are trying to link computers with the nervous system via implants—a precursor to the technology used by the cyborg assassins in *Angel.* For example, in 1997 scientists at the University of Tokyo attached a few motor neurons in a cockroach to a microprocessor and then sent artificial signals through electrodes. The cockroach was compelled to move wherever it was directed because of the electrical stimulation to the insect's motor cortex. Such devices are also similar to the chip the Initiative implants in Buffy's college boyfriend, Riley. We learn in "Primeval" (B-4) that the chip is tied directly to Riley's central nervous system through the thoracic nerve and can control his will.

Earlier in the season, the Initiative implants a different chip in Spike's brain to keep him from harming humans. Spike's chip is a behavioral modification device, but it still uses electricity to stimulate a specific part of the brain—in Spike's case, the region that processes pain—whenever he attempts to harm a human. Any attempt to do so results in a searing migraine so severe, not even vampires—who tend to get off on pain—can function. The chip must also be connected somehow via sensors to those regions that control violent emotion and motor function, since it apparently can tell when he is about to attack. Why the chip still enables Spike to hurt and kill demons is never satisfactorily explained. (Clearly a different region must be stimulated when the violent behavior is directed toward a nonhuman, but this makes no sense based on what we know of the brain and how it functions.)

Eventually Spike's chip degrades to a life-threatening degree, and the implant is removed ("The Killer in Me," B-7). This deterioration is analogous to the biocompatibility issues facing implantable silicon devices. Silicon tends to degrade quickly in the body, leading not only to chip malfunction, but also to infection. Warwick sidestepped the biocompatibility issue by encasing his electronic implants in glass tubes. Glass is chemically inert, nontoxic, and also doesn't block radio signals. But it merely substitutes one risk for another. There is always the chance that the tube could break and cause serious internal injuries.

When Team Angel takes over Wolfram & Hart ("Conviction," A-5), Gunn receives a special "brain upgrade," outfitting him with encyclopedic knowledge of both human and demon law, as well as the complete works of Gilbert and Sullivan, which are believed to enhance a lawyer's oratory skills in court. At the University of Southern California, scientists are working on a silicon chip implant that mimics the hippocampus, an area of the brain known for creating memories. The device resembles a crude prototype of Gunn's more mystical upgrade, although the USC researchers are still a long way from making such a microchip work in a human. Biocompatibility is the biggest hurdle and also the primary reason that the silicon retina currently under development by Optobionics—with an array of 5,000 microscopic solar cells to replace the eye's natural photoreceptors—has yet to undergo human clinical trials.

Warwick, for one, wouldn't find any of this preposterous. He believes that anything associated with a computer link could also be controllable through implants. It's possible that one day such an implant could hook directly into the brain, allowing humans unprecedented access to the world Internet memory bank. Even Willow's telepathic ability has a scientific counterpart. Warwick believes implant technology will one day enable humans to communicate across the Internet using thoughts alone, making human speech and language obsolete.

BATTERIES NOT INCLUDED

A lovesick Spike takes cybernetics one step further in "Intervention" (B-5). He bullies electronics whiz Warren Meers (later the leader of the Trio) into building him a robot version of Buffy, after the real Slayer rejects his amorous advances. For Spike, the Buffybot is a fantasy come to life. She is the ultimate battery-operated bimbo. To make her seem more like the real Buffy, the Buffybot has detailed program files on family, friends, demonology, weaponry, and martial arts.

After Buffy's untimely death in "The Gift" (B-5), the Buffybot is reprogrammed by Willow ("Bargaining," B-6). She passes for the Slayer to prevent Sunnydale from being overrun with demons, who might be drawn to the town by a Hellmouth that no longer has a guardian. The situation is far from ideal, but the Buffybot provides a convincing enough illusion—for a while, at least. Giles even finds himself training the robot as if she were human, instructing her on proper breathing and the concept of chi.

Robots are no longer the stuff of science fiction; indeed, they are almost commonplace, although few real-world robots truly resemble human beings. In Japan, robots are now used as receptionists, night watchmen, hospital workers, guides, babysitters, and pets. It's estimated that every household in Japan will own at least one robot by 2015. The designs are quite advanced. Alsok, a Japanese security guard company, has a line of robocops that detect and thwart intruders using sensors and paint guns. NEC's robotic babysitter can recognize individual children's faces and notify parents by cell phone in case of emergency. Cyber-receptionists are equipped with sophisticated voice-recognition technology allowing as many as 700 separate verbal responses. And Paro, a robotic baby seal, can recognize its master's voice, coos with delight when stroked, but squeals in pain if handled roughly.

In contrast, the United States has focused much of its investment in advanced robotics for military rather than consumer applications. For example, in 2005 the Army put a squad of remote-controlled robotic vehicles into the battlefield, armed with automatic weapons that fire when directed by radio signals. The military is also currently testing the first prototypes of a new unmanned ambulance. But the United States is not without its commercial applications: iRobot's Roomba is a robotic vacuum cleaner. There is even an annual RoboGames competition in which people design and build their own robots and pit them against other machines in a wide range of events.

However, all these technological marvels are little more than very sophisticated user interfaces. They do not exhibit true intelligence. Real-world scientists have built computers and robots that can convincingly mimic human behavior, like the Buffybot does, but they have yet to unlock the secrets of human emotion and consciousness. Electrical circuitry still has its limitations. Some scientists believe that all human thought is essentially algorithmic and can be broken down into a set of mathematical operations. They believe that we will one day be able to replicate the human mind and create a genuinely self-conscious robot capable of both thinking and feeling.

Depending on how you define it, "thinking" is already a reality for modern computers. In the 1940s, a British computer scientist named Alan Turing maintained that if a computer succeeded in fooling a human interrogator, placed in a separate room, into thinking it was human, then the machine could be said to be "thinking." This is known as the "Turing test," and for years it served as a working definition for artificial intelligence (AI). By this definition, the Buffybot would constitute AI. After all, her programming is extensive enough that she is able to give reasonably convincing responses in casual social settings.

But the standards for AI have changed substantially, and the question of what constitutes human "consciousness" has become increasingly controversial. For many scientists, the Turing test is not a measure of true consciousness, nor does a computer "think" in quite the same way that a human being does. The brain is often viewed as a massive parallel processor, but there is something innately inflexible about even the most advanced computer logic. This seems to be the predominant view in the Buffyverse. "The rules are the rules. Right is right and wrong is wrong. Why don't people see that?" demands Ted (B-2), a computer salesman who turns out to be a homicidal robot.

The Buffybot is no exception. Her programming can't quite capture the spontaneous wit of the real Slayer. She is inordinately fond of knock-knock jokes, and every time she stakes a vampire, she declares, "That'll put marzipan in your pie plate, Bingo!" There is also the occasional remnant of Spike adoration after she is reprogrammed, much to the grieving vampire's chagrin. Her lack of spontaneity also hurts her fighting ability. Preprogrammed moves, no matter how skillful, aren't sufficient to handle the unpredictability of an actual street fight. When a vampire catches the Buffybot off guard one night, the blow short-circuits her navigational system and she must return to Willow for repair.

Some scientists still believe that human thought and emotion can, at best, only be simulated by computers. A computer might seem intelligent, but it is not aware of what it is doing, having no real sense of self or consciousness. "I can't resist the sinister attraction of your cold yet muscular body," the Buffybot dutifully coos during a lovemaking session with Spike. But she destroys the illusion when she later asks, "Shall I start this program over?" The Buffybot is merely engaged in complex pattern recognition, analyzing input (the question or statement directed to her) and scanning her memory banks for the most appropriate response. She understands neither input nor output in the sense that a human being would. In much the same way, a mere machine with a preprogrammed set of responses cannot be considered truly intelligent.

This is the essence of the "Chinese room" parable proposed by John Searle, a professor of philosophy at the University of California, Berkeley. Place an English-speaking man inside an enclosed room with a rulebook that tells him how to respond to Chinese sentences that are pushed through a slot in the wall. To the observers, he appears to be fluent in Chinese, because his responses are given so fluently. But the man is merely processing requests. He doesn't understand either the phrases or his responses, any more than the Buffybot fully grasps the contextual meaning of her conversations with the various Scoobies.

Not everyone agrees with Searle's rather simplistic analogy. However, even Warwick the would-be cyborg has his doubts as to whether any machine could ever be truly sentient in the human sense. Rather, he believes that robots and computers will always perceive the world very differently from humans. This makes sense, since a computer's architecture is different from the neural structure of the brain. Most electronic systems are either analog (radios, TVs) or digital (microprocessers), but the brain is a combination of both. Sensory perception is generic analog input: Buffy feels a vampire's fist strike her face and the nerves register pain, sending a signal to the brain via chemical messengers. The brain digitally processes this information to draw a conclusion: Buffy has been hit. Neurons in the brain make digital decisions based on the collective analog data they receive as raw sensory input.

Among the most outspoken opponents to Searle's view is Daniel C. Dennett, a cognitive scientist at Tufts University, who believes that there is no intrinsic difference between the way humans think and AI. For Dennett, human consciousness isn't something that is "hard-wired" into the brain's machinery, but more closely resembles software programs that operate on its parallel processing hardware. Our consciousness has evolved over time. Much like the Buffybot, the human brain continuously processes incoming information, choosing appropriate responses and developing more complexity with each successive iteration. Not just AI, but artificial emotion might one day be possible, according to Rodney Brooks, whose lab at MIT is among the world's best in terms of developing cutting-edge robotics. Brooks believes that if robots can adapt, then they must be capable of evolving at least rudimentary emotions. He is best known for building Kismet, a robot capable of responding to social cues with convincing displays of emotion.

Adam is a cyborglike creature with equal parts human, demon, and electronic components, who terrorizes Sunnydale in Season 4 of *Buffy*. In the time-honored tradition of the sci-fi genre, Adam evolves beyond his programming and develops a mind of his own, killing his creator and running murderously amok. Adam is the perfect marriage between man and machine, far more than a simple brain-to-computer interface. He claims to be aware of every molecule in his being, more conscious than any human has ever been.

So is Adam the ultimate realization of AI? Not quite. Human emotion is still a bit beyond his comprehension. Love and friendship are intangible concepts that hold little meaning, and hence no value, for him. They might as well be magic. That's why he underestimates the "enjoining spell" the Scoobies perform to combine their respective strengths in Buffy so she can defeat him in "Primeval" (B-4). In the end, Adam is limited (and defeated) by the very electronic circuitry that gives him unprecedented cognitive abilities.

3

ROUGH MAGIC

Spellbound by the Laws of Thermodynamics

"This isn't 'Nam. This is bowling. There are rules."
—Walter, THE BIG LEBOWSKI

Like all Slayers before her, Buffy suffers an untimely death. In "The Gift" (B-5), she leaps into a vortex of mystical energy to stop yet another pending apocalypse, offering her own life in her sister Dawn's stead. The unnamed powers in the Buffyverse deem this a fair trade, and the apocalypse is duly averted. The Scoobies bring Buffy back to life three months later, thanks to Willow's increasingly potent magical powers ("Bargaining," B-6). Yet Buffy's resurrection comes with a nasty complication: the creation of a disembodied demon ("Afterlife," B-6).

Willow attributes the demon's presence to something called

thaumogenesis, a side effect to the powerful spell they cast. The Scoobies inadvertently created the demon. "The world doesn't like you getting something for free, and we asked for this huge gift: Buffy," Willow explains to the others. "So the world said, 'Fine, but if you have that, you have to take this, too.' And it made the demon." The demon's creation is temporary unless it kills Buffy, yet absolutely necessary. Magic always comes with a price tag attached, because opposing forces must be balanced, just like both sides of a mathematical equation. Nothing is ever free.

This is the most fundamental law of the Buffyverse, and it bears a striking resemblance to the conservation laws that dominate real-world physics. This is not mere coincidence. Willow unequivocally makes the connection between the two in "Get It Done" (B-7), when she declares, "Magic works on physics!" citing energy conservation by way of example. When Willow waves her fingers and utters an incantation to start a fire on the beach in "Buffy vs. Dracula" (B-5), she explains to her friends that it's all about balance: "You take energy from one place and bring it to another."

That's the essence of energy conservation. Extra influxes of energy, in theory, can account for many of the mystical phenomena in the Buffyverse, but since there is only a fixed amount available, the additional energy must be borrowed from somewhere else. Small wonder that Willow, the stellar science student, develops into such a powerful Wiccan. She understands that magic has its own set of rules that must be followed as rigorously as the laws of physics. There are loopholes and ingenious ways to "trick" those laws every now and then—the basis for all technological innovation—but even with magic, you can't cheat the universe. You have to play by the rules.

EQUIVALENT EXCHANGE

The Trio monitors Buffy's movements from the safety of a black van that is outfitted with all the latest high-tech surveillance equipment in "Life Serial" (B-6). To ward off discovery as Buffy is approaching the van, Jonathan disguises himself with a glamour, appearing to Buffy as a big red demon. She manages to punch him once before Jonathan beats a hasty retreat.

Once he has reverted to his smaller self, Jonathan confesses to Warren and Andrew how much the punch hurt: "I only looked big. I actually had the proportional strength of, uh, me." That's because in physics, all things conserve: matter, mass, momentum, energy, even charge. Each of these properties exists in a fixed, finite amount. Jonathan can spread his existing atoms and molecules over a greater surface area, but he can't gain extra mass and energy—and therefore greater size and strength—from a mere glamour. All he achieves is the illusion thereof. His overall mass and energy remain the same.

The notion of conservation of matter is one of the oldest laws of physics, with roots in the alchemical principle of equivalent exchange. The term "alchemy" is derived from *chemia*, the Greek for "the Egyptian art," since the practice was believed to have originated in ancient Egypt, despite having flourished, particularly during the Middle Ages, in Europe. The precursor to modern-day chemistry, alchemy was an ancient, arcane art devoted chiefly to discovering a substance that would transmute more common metals into gold or silver using something called a "Philosopher's Stone." Some alchemists believed that grinding up the stone and boiling it in water produced an Elixir of Life that cured all ailments and enabled the drinker to live forever. Neither objective was ever achieved, and alchemy was ultimately discredited by the late seventeenth century.

For all their association with magic, alchemists did become quite adept at preparing various chemicals and refining the associated processes—including distillation, condensation, evaporation, and combustion. And they firmly believed that in order to gain an objective, something of equal value must be lost. This is the principle of equivalent exchange. Gold could not be created out of nothing. It had to be transmuted from other metals—usually lead, but certainly something equally dense and heavy. We see this in chemical reactions all the time: The total weight of the chemicals remains unchanged because the total number of atoms involved in the reactions doesn't change. There is no detectable increase or decrease in the quantity of matter in the average chemical reaction, even if it sometimes seems that way. For instance, if Willow weighs a piece of copper metal and then heats it in chemistry class, its molecules will mix with the oxygen molecules in the air. When Willow then reweighs the copper, she will find that it has greater mass than before. The oxygen in the air combines with the copper, adding their masses together. Once Willow takes into account the combined masses of the original oxygen molecules and the original piece of copper, her total will be equivalent to the mass of the copper after the reactions take place.

Conservation of matter might explain why some sort of exchange is required whenever someone in the Buffyverse casts a spell—basically a chemical reaction with mystical trappings. You can't get something from nothing, and the greater the boon requested, the greater the sacrifice required. In "Reptile Boy" (B-2), fraternity brothers periodically sacrifice three young women to a giant demon snake, who grants them continued wealth and power in return. Demon puppets hijack a children's TV show in "Smile Time" (A-5) and suck the innocence out of their young viewers in lieu of the usual advertising revenues. As the head demon tells one little boy, "You know 'Smile Time' isn't free!" And thaumogenesis rears its ugly head again in "Superstar" (B-4). Jonathan casts a spell transforming himself into a celebrity paragon—the best at everything. The catch is that the spell must also create an equal and opposite monstrous creature embodying the worst of everything. The two are inextricably interlinked: killing the monster breaks the spell, and Jonathan reverts to his former nebbishy self. Phrased in more mathematical terms, both sides of the magical "equation" must balance out.

Conservation laws are so tightly woven into the fabric of the Buffyverse that the writers will go to great lengths at times to ensure that balance is maintained in the story arc, not just in individual episodes, but from season to season. Early on in *Buffy*, Angel stakes his own sire, Darla ("Angel," B-1), who is resurrected on *Angel* as a human by Wolfram & Hart ("To Shanshu, in L.A.," A-1). In Season 2, Darla finds out she is dying, and Angel nobly undergoes a series of trials to win back her life ("The Trial," A-2). But Darla has already been given a second chance by mystical means; his request can't be granted. Ultimately Darla is sired all over again as a vampire by her own granddaughter, Drusilla.

So the Powers That Be owe Angel a life. There is a credit carried on the books of fate, until they find a rather ingenious way to repay him. Vampires reproduce through the process of siring, not through normal human conception. Yet we discover in Season 3 that the newly sired Darla has mysteriously conceived a son after one passionate night with Angel ("Offspring," A-3). The Powers have repaid their debt to Angel with this new life; balance is restored.

INSIDER TRADING

This balance of life-for-life is a recurring theme in the Buffyverse, although the required exchange occurs in many different ways. In "Get It Done" (B-7), Buffy jumps through a portal into another dimension, unleashing a demon from that other world in her place. To get her back, the Scoobies must make a similar exchange: the demon for the Slayer. But first they need to reopen the portal, and this requires energy. Anya uses the vocabulary of science while brainstorming with Willow over the best way to do it, insisting that they need some kind of catalyst to start the spell. This is not entirely sufficient. They also need an appropriate conduit for the transfer of energy.

Clearly, in the thousand years she spent as a vengeance demon, Anya picked up a little physics. What she is describing bears a strong resemblance to how a battery produces energy. Whenever one type of matter converts into another, as in a chemical reaction, the change is accompanied by the conversion of one form of energy into another. A battery has two ends, called terminals—one with a negative charge and one with a positive charge. Electrons congregate on the negative terminal. Connect a wire (the conduit) between the two terminals, and the electrons will flow from the negative to the positive end as quickly as they can. Connecting the battery starts the flow of electrons, catalyzing a series of chemical reactions inside the battery to create even more electrons. Once all the chemicals have been used up and converted into energy, the battery goes dead.

Energy transformations don't necessarily involve chemical reactions. There are many different kinds of energy that can change other kinds. For example, burning coal in power plants produces electricity by converting thermal energy (heat) into mechanical energy in a turbine. Electrical energy can change into mechanical energy. And an electrical generator converts mechanical energy into electrical energy, which can then be used to power most of modern technology. But all types essentially depend on the conversion of potential into kinetic energy.

To hold the portal open long enough for the Slayer/demon exchange to take place, Willow must borrow energy from the people around her—in this case, her new girlfriend, Kennedy, who is a Potential Slayer. "You were the most powerful person nearby, and that's how it works," Willow ruefully explains when Kennedy understandably feels violated. Just as Kennedy is waiting to be activated as a Slayer, potential energy (energy of position) is waiting to be converted into kinetic energy (energy of motion).

Potential energy can take many forms. The electrons that congregate on a battery's negative terminal are storing potential energy, which is converted into kinetic energy when the two terminals are connected by a conduit, and the electrons begin to flow to the positive terminal. When Buffy loads a stake into her crossbow in the cemetery to take out a vampire prowling for a victim, she increases the weapon's potential energy by winding the release mechanism, which is then converted into kinetic energy when the stake is released. And the higher an object is lifted from the ground, the more potential energy it gains, and the more kinetic energy results when the object finally drops. The total energy of any given object is simply its potential and kinetic energy added together.

That energy is useless, however, unless it can be harnessed to perform some function. A battery is usually connected to some kind of "load": That is, its energy is harnessed to operate a flashlight or a radio, for example. The energy produces work. How much work a given object is capable of performing is precisely equal to its kinetic energy. Work is defined in physics as a force applied over a given distance ($W = fd$). When Spike lugs the now-deceased demon back to the Summers house to make the exchange, he is performing work. The heavier the demon is, and/or the greater the distance Spike carries it, the more energy Spike expends to do the work.

Work is related to Newton's second law of motion, which states that force equals mass times acceleration ($F = ma$). The more force one applies to an object, the more it will accelerate and the farther it will travel, but this is balanced against Newton's first law. All objects possess an intrinsic property known as inertia—the amount of force required to get them to move or change direction while in motion. The more mass an object has, the more force is required to get it moving, and the more slowly the object will accelerate. Taken together, this means that Spike will expend the same amount of energy—and produce the same amount of work—carrying the heavy demon a short distance, as he would carrying a feather over a very long distance.

The concept of work in physics can give rise to some decidedly counterintuitive notions. Angel might take the elevator to the topmost floor of Wolfram & Hart's headquarters, then ride back down feeling as though he's accomplished something: moving from point A to point B and back again. But from a physicist's perspective, nothing has been done. Angel gains potential energy as the elevator rises because he increases his distance from the earth's center of gravity, but he loses potential energy as the elevator descends, and his kinetic energy increases. The two cancel each other out, and the amount of work produced is zero, unless Angel finds some way to harness the kinetic energy as he descends.

There are lots of clever ways to do this. For instance, aspiring high school magician Mark—actually a demon in human disguise—employs a guillotine as part of his magic act for the

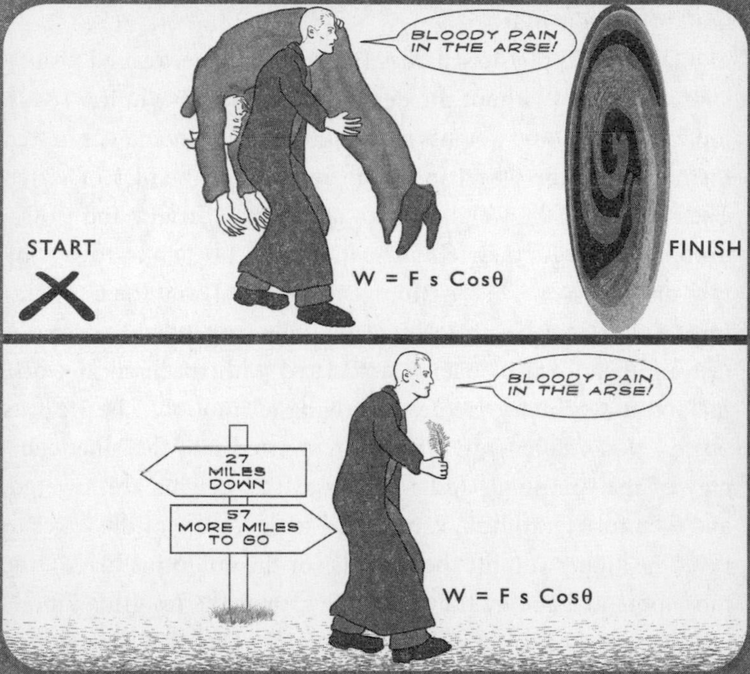
BLOODY PAIN IN THE ARSE!
START
FINISH
W = F s Cosθ
BLOODY PAIN IN THE ARSE!
27 MILES DOWN
57 MORE MILES TO GO
W = F s Cosθ

Sunnydale High School talent(less) competition in "Puppet Show" (B-1). When the blade is raised to the top of the contraption's frame with a rope, it stores potential energy, which is converted into kinetic energy when the rope is released and the blade plummets downward. (In a neat little physics-based aside, Giles explains to Mark how to calculate the velocity of the guillotine blade by calibrating the counterweights.) This kinetic energy is then harnessed to perform mechanical work: chopping off the head of Demon-Mark.

The British physicist James Prescott Joule performed a similar experiment (without the demonic decapitation) in the 1840s. Joule attached some weights to strings and pulleys, and connected them to a paddle wheel inside an insulated container of water. Then he raised the weights to an appropriate height and slowly dropped them. As they fell, the paddle wheel began to turn, stirring up the water. This friction generated heat, and the temperature of the water began to increase. Joule concluded that heat is just another form of energy associated with mechanical work, making it a primary factor underlying all motion. The weights possess potential energy, which is converted into the kinetic energy of the spinning paddle wheels as the weights are lowered, and then turns into heat, causing the temperature of the water to rise. The higher you lift the weights (or the guillotine blade), the more potential energy they gain, the greater the resulting kinetic energy produced as they fall, and the more heat that is generated by the spinning paddle wheels (or the more force the blade exerts against the neck of the unfortunate Mark).

Thus, just like the First, energy can take any number of forms. But the laws of conservation still hold. Regardless of the type of conversion taking place, in the end there must be the same total amount of energy after the exchange as there was before. Case in point: when Spike crashes Giles's car in "A New Man"

(B-4), the vehicle possesses a certain amount of energy, depending on the speed at which it is traveling when it hits the brick wall. Most of that energy of motion is converted into other forms of energy upon impact: the sound of the crash, heat energy from the friction resulting from the collision, the kinetic energy of flying brick fragments and the crumpling of the car body. Yet if we were to add the total energy of the various types together, it would be the same as the energy of the car just before the crash.

THE TIES THAT BIND

Energy isn't the only quantity that's conserved; mass is subject to the same principle. The townspeople of Sunnydale are whipped into a murderous frenzy in "Gingerbread" (B-3) by the apparent death of two young children with occult marks on their bodies, triggering their own version of the Salem witch trials. Willow, Buffy, and a teenage witch named Amy soon find themselves tied to stakes over piles of confiscated books, to be burned in revenge for the senseless killings. As the flames begin to rise, Amy escapes her imminent death by turning herself into a rat, disappearing into a pile of clothing and scurrying out of danger.

There's just one problem with Amy's ingenious escape. Her rat self has significantly less mass than the human Amy, and mass is always conserved. It can't just vanish into nothing, yet it clearly hasn't been turned into another material object. So where did the missing mass go? This isn't a new problem in physics. A similar apparent violation of conservation laws occurs with the elements in the periodic table. The atomic number of an atom denotes how many protons it has (and therefore which element it is), while its atomic weight includes both protons and neutrons. For elements lighter than iron, the atomic weight is smaller than the mass of the protons and neutrons together, while for elements heavier than iron, the atomic weight is a little larger than the total mass of the neutrons plus protons. So the lighter elements are missing mass, while the heavier ones have extra mass, which on the surface appears to violate conservation laws. Scientists pondered the mystery for decades.

Albert Einstein solved the puzzle in 1905. It turns out that mass can be converted into energy, and vice versa, in much the same way that different types of energy can convert back and forth. Mass and energy are equivalent, interchangeable. The extra mass in the heavier elements comes from the extra binding energy needed to hold the protons and neutrons together to form the nucleus—a bond much tighter than the ropes that bind Buffy and her friends to their respective stakes.

We've already seen that an atom can absorb energy, causing its electrons to jump to higher orbitals. This extra energy is then emitted as radiation. Something similar happens when an atom loses protons or neutrons: This changes its atomic weight, and the missing mass is radiated as energy. As an atom radiates energy, it gets lighter, and disintegrates into an isotope of the same element with the same number of protons but fewer neutrons. Or it could decay into a different element of lesser weight if it loses protons. Heavy elements have lots of energy to spare. The heavier they are, the more extra energy they have, which is why really heavy elements like uranium are so radioactive, emitting high-energy gamma rays until they decay into stable elements.

So matter doesn't disappear, and energy doesn't spring out of nowhere. Instead, energy and mass can change into each other. Einstein summed up this principle in his most famous equation,

$E = mc^2$. "E" is energy, balanced against "m" for mass, while "c^2" is the exchange rate between them: the square of the speed of light. This is a very large number, which explains why a small amount of matter can produce a tremendous amount of energy—far more than a normal chemical reaction.

Amy's transformation is accompanied by a brilliant light display, and light is a form of radiation energy. So the missing Amy-mass was released as radiant energy. Exactly how much energy are we talking about? The average human body contains roughly 10^{28} atoms. Assuming that Amy weighs roughly 120 pounds,* converting most of her body mass into energy on such a short time scale would release the radiation equivalent of more than one thousand 1-megaton hydrogen bombs. This should pose a serious problem for anyone who happens to be in the vicinity during the transformation. The assembled crowd of would-be witch burners ought to have been burned to a crisp, or at least have suffered burns, lesions, nausea, and the occasional cancerous tumor. Yet somehow none of them suffers any ill effects.

The high exchange rate provided by the squared speed of light offers a possible explanation for why magic in the Buffyverse requires the sacrifice of some material substance. The mass of that substance is converted into energy, which is then harnessed to perform the spell. How much energy can be harvested depends on what is being sacrificed. There appears to be an "energy hierarchy." Inanimate ingredients, for example, seem to possess less intrinsic energy (for magical purposes) than an animal, while a human possesses more energy than an animal, and a demon—or a Slayer—possesses more than the average human. "Higher Beings"

*Disclaimer: all stated weights are estimates only, rounded off for easier calculation.

in the Buffyverse are most powerful of all. They seem to be able to tap at will into a mysterious, vast reservoir of mystical energy that is inaccessible to lesser beings, making them capable of some truly astonishing feats.

When Anya returns to her vengeance-demon ways, she grants a particularly brutal wish in "Selfless" (B-7) that leaves a dozen frat boys dead with their hearts ripped out. Wracked with guilt, Anya asks her demon boss, D'Hoffryn, to undo the spell and restore their lives. "You know how this works. The proverbial scales must balance," he tells her. A sufficient influx of energy is needed to restore twelve human lives, and the best energy source for that is the life and soul of a vengeance demon. Anya offers herself in exchange for undoing the spell, although in a cruel twist, D'Hoffryn takes the life of her best friend, fellow vengeance demon Halfrek, instead. Halfrek disintegrates in a blinding flash, her matter converting into the energy needed to restore the lives of the dead young men.

Einstein's equation works both ways. Energy can also convert into mass, although such a feat is much more difficult. Unlike nuclear decay, it doesn't occur naturally, because it requires large amounts of energy at very high temperatures. The more mass an object has, the more energy that is required to produce it. In Season 3 of *Buffy*, we learn that Mayor Wilkins wishes to be transformed into a giant snake demon on graduation day—a process known as Ascension. But he needs an enormous influx of energy in order to add that much mass to his body. He acquires it by ingesting a box full of icky black spiders that contain heaps of mystical energy.

By way of real-world comparison, it takes a lot of energy just to create a top quark (the most massive particle in the quark family) in a particle accelerator. Protons are accelerated to speeds very close to the speed of light, and when they collide with antiprotons, the impact releases a burst of energy on the order of 1.8×10^{12} electron volts. This is about the same amount of energy one would need to do a single push-up. That might not sound like much, but it becomes much more impressive when we consider the relative masses of a quark and Mayor Wilkins: All that energy is concentrated into a tiny subatomic particle instead of an adult-sized human being. Mayor Wilkins would have to clock in at a breakneck pace of roughly two miles per second during a morning run to produce a comparable amount of kinetic energy.

It stands to reason that the more mass an object has, the more energy will be required to produce it. The top quark is 58,000 times heavier than the lightest "up" quark, so it takes 58,000 times more energy to create a top quark than it does to create an up quark. Remember the classic carnival game where one swings a sledgehammer to hit a target, which then causes a weight to fly up and strike a bell? The difference in energy required to make an up quark versus a top quark is roughly the same as the difference between swinging the sledgehammer hard enough to make the weight rise a measly three feet and swinging it hard enough to send the weight to the top of the earth's stratosphere.

There's an even greater difference between the energy required to create a top quark and that needed to create a giant snake-demon. A top quark's mass, expressed in units of energy, is 175 GeV/c^2. But when that mass is converted into macroscale units, it becomes minuscule: 0.00000000000000000000011 ounces (yes, that's 22 zeros between the decimal and the 1). Mayor Wilkins weighs roughly 170 pounds, considerably more than the top quark. The demon he turns into is about the size of a brontosaurus, and would weigh more than 30 tons. To produce sufficient mass to make such a transformation would require the same amount of energy that is needed to power 75 trillion American homes for an entire year. That adds up to a lot of magic spiders. No wonder the mayor is up all night consuming them.

This also may be the reason Willow has so much trouble

turning Amy back into her human form. Though it would take a far smaller influx of energy to reverse Amy's transformation than it would to turn Mayor Wilkins into a giant snake, the amount of energy Willow would need to do so is still, as physicists like to say, nontrivial: enough wattage to power 150 million American homes for a year. Poor Amy languishes in her Habitrail cage for three more seasons until Willow finds a solution in "Smashed" (B-6). It turns out to be ridiculously simple. Willow literally plucks the required incantation from out of thin air, and within minutes Amy returns to human form. There is an unsatisfying whiff of deus ex machina to Willow's achievement, perhaps because there is no attempt to explain where she acquired the extra energy to restore Amy's missing mass—a rare omission by the series' writers. Still, one could argue that she has only just figured out how to tap into the huge energy reservoir that seems to fuel so much mystical activity in the Buffyverse.

Alas, even that argument doesn't hold when considering the case of recorporealizing Ghost-Spike, a process that requires an equally massive surge of mystical dark energy as a catalyst—"the equivalent of nuclear evil," according to Fred. She builds a machine capable of achieving the feat in "Hellbound" (A-5), and locates a prime source of concentrated dark energy: the mystical conduit connecting Wolfram & Hart to all the other dimensions. Unfortunately, another ghostly being is recorporealized instead.

The solution that eventually transpires is far from satisfying. Fred admirably sticks to the laws of physics (as they exist in the Buffyverse) when attempting to solve the dilemma. So it's difficult not to feel a bit cheated, from a physics perspective, when Ghost-Spike simply opens a mysterious package in "Destiny" (A-5) and is hit by a flash of light. This instantly recorporealizes him. The flash had about as much energy as the average flash bulb, yet somehow it did the trick. Fred—a highly trained scientist who is nonetheless open to the workings of mystical phenomena—understandably finds this hard to swallow. The bitter truth is that sometimes there simply is no way to undo the damage that's been done, even in a magical world like the Buffyverse.

IRREVERSIBLE YOU

In "Villains" (B-6), a grief-stricken Willow turns to black magic to seek revenge against the Trio for the death of her (pre-Kennedy) girlfriend, Tara. She siphons all the mystical energy out of the dark arts books in the Magic Box occult shop. But halfway through a high-speed chase down the Interstate, Willow suddenly collapses. All her borrowed power has been used up. As a greatly relieved Jonathan puts it, "Running that hot for that long, it's only a matter of time before you need to get juiced up again."

Thanks to energy conservation, no machine or living creature can produce more energy than it consumes. But there's a corollary. Theoretically, a well-designed machine should be able to run indefinitely on a fixed supply of energy, provided it recycles the energy it takes in to produce new energy with 100 percent efficiency. Unfortunately, the most efficient machines humans can build still lose some heat energy as they operate. There is no such thing as a perpetual motion machine—not one that produces energy from nothing, and not one that produces exactly as much energy as it consumes.

The reasoning is simple. Machines have moving parts, and those parts generate friction and heat. While heat, like other forms of energy, can be harnessed to perform useful work, a small amount will always dissipate into the atmosphere and be lost forever. So no machine can run indefinitely without constantly replenishing the lost energy. That's why Willow can't run forever on the fixed supply of energy she gains from the books. When her power supply runs low, Willow must find a way to recharge: first by way of Rack—a shady dealer in black-market magic—and later by siphoning extra power from Giles when he arrives to help stop her.

Physicists attribute this constant need for replenishment to a phenomenon called entropy. Technically, entropy is a measure of how much disorder there is in a given system, and hence, how much useful work that system can produce. The more order that is present, the more energy is available for work. The greater the disorder, the less energy is available for work. The mystical books in the Magic Box, neatly sorted and stacked, are very well ordered. In Willow's confrontation with Buffy and Giles, the books are blown apart, their pages strewn randomly around the store. They become disordered. There is no way to restore their original order without bringing in additional energy—in this case, Anya, who spends several months cleaning up the detritus of Willow's vengeful rage. The text contained in the books is also highly ordered: letters form words, words form sentences, and so on, albeit in strange, ancient languages. Mystical languages are potent things in the Buffyverse, and this high state of order gives the books a great deal of potential energy. Willow releases that energy when she drains the books of their words, and hence their power.

As always, there is a catch: All forms of energy are not created equal. Heat is much more disordered (high entropy) than, say, potential energy, which is highly ordered (low entropy), so the energy of the latter can be more easily harvested for work. Although the various types of energy are interchangeable, heat only flows in one direction. Burning coal produces heat, which can be harnessed to perform work, but once that coal has been reduced to ashes, the heat energy it produced is lost forever.

Physicists call this an "irreversible process," and it is a direct result of entropy. A cooler body can't pass heat to a hotter one any more than a hot cup of coffee can become even hotter while sitting in a colder room, or melted ice cream can refreeze of its own accord. Just the reverse happens: Coffee cools, the ice cream melts into liquid. Things, in general, decay, unless some outside force intervenes to counter the process. We see the same thing in the Buffyverse: Vampires, being technically dead, are cold and unable to generate their own heat energy. They can leach heat from their surrounding environment, but only if that environment is hotter than they are. They can't draw heat energy from a cold room any more than that cup of coffee can.

That's why it takes so much energy to produce a small amount of mass. Although $E = mc^2$ technically allows for conversion in either direction, the exchange rate clearly favors mass-to-energy conversions, which release heat as part of the spectrum of emitted radiation. A certain amount of the heat energy produced will dissipate and be lost, no matter how efficiently one tries to harness it. Willow succeeded in turning Rat-Amy back into her former self, but if the laws of thermodynamics are strictly followed, some small portion of Amy's original mass should be lost forever. Since Amy was a bit self-conscious about her weight when human, perhaps she won't mind the loss.

Of course, this doesn't take into account other forces that might come into play, such as gravity or electromagnetism, which can add energy to a system. It only applies to closed systems that don't interact with the outside world. Consider the refrigerator in the Summers kitchen. Refrigerators produce useful work in the form of cooling by drawing energy from differences in temperature

GODDESS HECATE, WORK THY WILL!
E=mc²
1eV/c²=1.783 x 10⁻³⁶Kg
S = k 1n Ω ?

and pressure. A gas (usually ammonia or freon) is pressurized in a chamber, which causes it to heat up. The coils on the back of a refrigerator dissipate this heat, and the gas cools and condenses into a liquid. It undergoes a phase transition. The liquid then flows to a second, low-pressure chamber. The change in pressure causes the liquid to boil and vaporize, dropping its temperature and making the inside of the refrigerator cold. The cold gas is then sucked back into the first chamber, and the entire cycle repeats ad infinitum. Conceptually, a refrigerator is a closed system, since the gas is continually recycled. But thanks to entropy, it will irrevocably lose a small amount of energy every time it completes a cycle. That's why all refrigerators need to bring in a constant influx of electricity to continue operating. If Buffy were to unplug the refrigerator, the interior would gradually become warmer as heat seeped into the cooler reservoir—spoiling any food (or pig's blood) left inside.

Therein lies the secret to how life is possible. Living organisms are constantly taking in energy in the form of food and oxygen, which are turned into mechanical energy and heat. Even in the Buffyverse, no matter how exotic the creature in question, the monsters all have one feature in common: the need for a constant replenishment of energy, whether it be flesh and blood, or more intangible things like fear, paranoia, or sexual energy. Once a creature stops taking in nourishment, its metabolism shuts down and the decay process sets in. Vampires are technically dead, but they receive energy from the demon inhabiting the human vessel, and the energy is supplemented with regular infusions of blood to ward off physical decay. When Giles asks Spike what happens to vampires who don't feed ("Pangs," B-4), Spike replies, "Living skeletons, mate."

Mathematically speaking, irreversibility isn't quite as absolute as it appears. In the nineteenth century, an Austrian physicist named Ludwig Boltzmann redefined entropy as more of a statistical law applying to molecules en masse, not individually. A pint of ice cream contains billions of atoms or molecules. When frozen, those atoms are arranged in a precise, highly ordered, crystalline structure. As the ice cream heats up and begins to melt, that order gradually disappears until the ice cream turns into a liquid. There is only one way for the atoms to be arranged in order to form a solid, but several combinations for them to form a liquid, and still more besides for them to form a gas (vapor). So it is merely highly unlikely—but not impossible—that the molecules in the melted ice cream will return to their original, orderly frozen state. The same goes for the randomly strewn pages of the mystical books in the Magic Box: There is the tiniest statistical possibility that those pages will reorganize themselves back into their original order. One would just have to wait long enough—an eternity or so—for this to happen.

If one happens to live in the Buffyverse, one can perform a magical ritual to acquire a burst of additional energy to speed things along. When a vamp gets "dusted," the creature is reduced to billions of particles of ash—a state of extremely high entropy. It is well-nigh impossible to restore the order that has been lost, at least not without an appropriate sacrifice to achieve a huge influx of energy. For example, five vampires must be sacrificed in Wolfram & Hart's ritual in order to generate sufficient mystical energy to bring Darla back as a mere human.

Similarly, vampires must sacrifice their human souls in exchange for immortality and all the superpowers that accompany the transformation. Angel has regained his soul by means of a gypsy curse, but he briefly loses it again on two occasions: once in Season 2 of *Buffy*, and again in Season 4 of *Angel*. In both cases, his disembodied soul is depicted visually as a vapor, so we can liken souls in the Buffyverse to a collection of molecules in a gas.

Entropy dictates that all systems tend toward a state of equilibrium, in which there is no net flow of energy. Mix hot water with cold water, and the atoms will mingle; heat will dissipate among them until they are all at the same temperature. Gas molecules are held together by their container, but when the gas is released, the molecules disperse outward. They have more space in which to jostle, many more atoms with which to collide in the surrounding atmosphere, so they will naturally drift to higher and higher entropy until all the molecules are uniformly spread out into the larger space.

Angel's soul (the gas) left his body (the container) when he became a vampire, but it still existed, floating about the Buffyverse in a disembodied state, virtually indistinguishable from all the other atoms in the atmosphere. It is highly unlikely that a gas will re-coalesce into its former container, and just as unlikely that Angel's soul will re-coalesce back inside his body. Neither could one toss a glass of water into the ocean and get the exact same glass of water back, right down to every individual molecule. That gypsy curse provides sufficient mystical energy to improve the odds and "reverse" entropy in Angel's case—the last two times courtesy of Willow, the only living person powerful enough to work the complicated spell. The same goes for Spike, who literally buys back his soul through a series of torturous trials at the end of Season 6 of *Buffy*. So the occasional loophole is possible in the Buffyverse.

There are still inherent limitations. The Scoobies succeed in bringing Buffy back to life, because her death wasn't natural. She was killed by mystical energy, giving them a better chance at magically reversing the death-and-decay process. But when Willow calls on Osiris again to restore Tara's life ("Villains," B-6), her request is refused. Tara's death is part of the natural (entropic) order of things, and cannot be undone.

Resurrection spells in general are considered taboo. On the rare occasions when they are performed, they usually go horribly wrong. We can refreeze ice cream after it melts by putting it back in the freezer—adding energy to reverse the melting process—but the ice cream's consistency is never quite the same. When Dawn and Spike visit a demon for a resurrection spell to bring back Dawn's and Buffy's mother, Joyce, from the grave ("Forever," B-5), they learn that it is possible to bring back some semblance of the person who was lost, but there is always an essential part that can't be regained. Just as ice cream that melts and then refreezes will irrevocably alter its consistency, some energy is always irretrievably lost in the transformation from living to dead and back again. Angel and Spike regain their souls, but not their full humanity. They are still undead creatures with mystical superpowers who need to consume blood daily. Even Buffy isn't quite the same after her resurrection. Apart from the psychological trauma, the process of funneling her "essence" back into her body somehow alters her surface molecular structure, according to Tara, who compares it to a "deep tropical cellular tan."

When Fred is killed and Illyria takes over her body, Angel assumes that there is a way to get her back, because Fred's soul should still exist. "Death doesn't have to be the end—not for us!" he declares in "Shells" (A-5). Since entropy is a statistical law, the remaining members of Team Angel could, in theory, restore Fred's soul to her body, just as Angel's soul was restored. They just need to know the exact positions and speeds of each and every molecule. The resources of Wolfram & Hart are considerable, far greater than our own, so it's conceivable that despite the inherent difficulty of performing such a calculation, their scientists might be able to determine precisely where each molecule of Fred's soul will be at a given point in time, and supply sufficient energy to return them to their former state. So there is the tiniest glimmer of hope for Team Angel. Even without mystical intervention, given enough time, there is an almost infinitely small chance that the dissipated molecules of Fred's soul will all find their way back to her former body of their own accord and reorganize themselves into the proper order.

Alas, Fred's death turns out to be a truly irreversible process. Her soul wasn't merely released from its vessel, it was consumed completely in the transformation process in a kind of nuclear reaction that fused Illyria's essence to Fred's human shell. Once heat energy has been spent, it is lost forever. Fred's soul was consumed, so it can't be regained. Everything that Fred once was is now part of Illyria. So Angel was wrong: Sometimes death does have to be the end. As Buffy tells a grieving Dawn in the wake of Tara's demise, "There are limits to what we can do. There should be."

NO EXIT

Those limits are what make life so frustrating and yet so precious. Buffy ultimately defeats the disembodied demon that was created as the price for bringing her back in "Afterlife" (B-6). But her return to the land of the living carries one other unexpected twist. Willow is a little nonplussed to find that Buffy isn't exactly gushing with gratitude toward the Scoobies. Buffy confesses to Spike that she had been in a sort of heaven. Mourning her loss of peace, Buffy concludes, "This is hell."

If hell is defined as the absence of hope, then Buffy may have a point. Consider the situation she faces. Even with all that extra mystical energy floating around, nobody in the Buffyverse—

human, demon, or hell god—can get something for nothing, and any victory is fleeting at best. Power always comes at a price, and magic spells always have consequences. Matters aren't much better in the real world. Thermodynamics is one big buzzkill, continually limiting our soaring technological aspirations with its pesky scientific fine print. Small wonder that physicists occasionally refer to them as the laws of "thermogoddamnics."

The physicist and author C. P. Snow drew an analogy between those laws and the house rules of a gambling casino. It is impossible to win outright, he said, because matter and energy are conserved. At best, a gambler might break even. The same goes for the Buffyverse. While the Slayer line was established to counter the spread of evil, Buffy can never completely prevail. She can't single-handedly wipe out every demon in every remote corner of the globe. Even backed by an army of Slayers, she can't destroy the First entirely, because evil has always existed. The best she can hope for in her battle against the forces of darkness is to hold things to a draw.

Yet in the great cosmic casino, you can't break even, either. The game is always rigged in the house's favor. Entropy dictates that there is always an increase in disorder, making it impossible for a closed system to return to a higher energy state. In the Buffyverse, disorder can equate metaphorically with evil, and evil keeps upping the ante. Buffy moves from fighting vampires and garden-variety demons to battling Adam (the ultimate cyborg), Glory (an exiled hell god), and the First. Her opponents keep multiplying and getting tougher to beat, and Buffy understandably finds this disheartening. "I battle evil. But I don't really win," Buffy tells Angel in "Gingerbread" (B-3), after her mother dismisses her demon-slaying efforts as fruitless. "The bad keeps coming back and getting stronger."

That's thermodynamics in a nutshell: It's the ultimate losing battle. Boltzmann, wrestling with the mind-numbing statistics of entropy, certainly understood this. After years of fighting a seemingly losing battle for acceptance of his ideas by the scientific establishment, he committed suicide in 1906 at the age of seventy, while vacationing with his family in Italy—just a short while before his theories were proved correct. His now-famous equation for entropy is engraved on his tombstone. (Buffy's own epitaph eschews mathematics in favor of a more fitting tribute: "She saved the world a lot.")

As if things weren't depressing enough, there's no getting out of the casino. We have no choice but to play a losing game, because the universe never really stops running. All the atoms that make up matter vibrate to some degree; how fast they vibrate depends on heat. The hotter the atoms, the faster they vibrate, and vice versa. Absolute zero is the theoretical point at which atoms become so cold, they cease to move entirely and exist in a state of absolute rest. "Time didn't mean anything," Buffy tells Spike when describing the heavenly dimension she inhabited after dying. Time doesn't mean anything to atoms at absolute zero either. If there is no movement, and hence no change, then there is no way to mark the passage of time.

For better or worse, the laws of physics conspire to ensure that nothing ever reaches absolute zero, technically defined as minus 495 degrees Fahrenheit, or 0 degrees Kelvin. The atoms in something as cold as liquid nitrogen still vibrate and generate some heat, despite their very low temperature. Even the coldest depths of outer space hover about 3 degrees above absolute zero. That's because space is filled with a uniform radiation field, the afterglow of the big bang, called the cosmic microwave background radiation. It ensures that temperatures in space never drop below 3 degrees Kelvin.

Physicists have come very close to reaching absolute zero in a lab environment. In 1995, scientists at JILA/NIST in Boulder, Colorado, managed to cool a small sample of cesium atoms to a mere few billionths of a degree above absolute zero, achieving a new state of matter known as a Bose-Einstein condensate (BEC). To do this, they essentially wove a "web" out of infrared laser beams. The laser beams bombarded the collected atoms with a steady stream of photons. Normally this would make the atoms hotter, because the atoms would absorb the extra energy. But certain wavelengths of light will bounce off the atoms instead, carrying away more energy than when the photons hit them. The atoms slow down as they cool. Time doesn't quite stop in a BEC, but to the atoms, it must certainly seem that way.

Should scientists achieve absolute zero one day, it's doubtful that it could be sustained for long. The same goes for Buffy. She attains a static state of absolute bliss after five brutal years of battling evil. But she isn't allowed to hover there indefinitely. After all, what is the Buffyverse without Buffy? Even in death, it seems, she can't escape her destiny. Once she's been resurrected, the vicious cycle starts all over again, with no end (or rest) in sight. In that respect, the "house rules" of the Buffyverse are every bit as absolute and unyielding as our own. No wonder Buffy is so depressed about being back in Sunnydale. She can't win; she can't break even—and she can't get out of the game.

4

THE PHYSICS OF THE FIGHT

Mass, Momentum, and the Martial Arts

"Ooooh . . . You think you're so cool, with your karate and your childlike reflexes."
—Roy O'Bannon, SHANGHAI KNIGHTS

Buffy butts heads in "Phases" (B-2) with a chauvinistic, rogue werewolf hunter who snidely derides her physical abilities. The boys at Buffy's high school who've had the misfortune to tangle with her know better, as do the various and sundry vampires and demons who cross her path. Buffy might be a little blond girl who looks like she could barely lift a crossbow, let alone fire one with deadly accuracy, but she is far from the stereotypical damsel in distress, and therein lies a large part of her appeal.

The plight of the Slayer is a poignant one, marked by loneliness, isolation, constant violence, and innumerable injuries, not to mention a sharply abbreviated life span: Most Slayers do not live past the age of twenty. But there are some consolations, not the least of which is a vast source of mystical Slayer strength and preternatural agility. But Buffy does not rely solely on her mystical gifts. She's a well-trained fighter who logs relentless hours of physically grueling training to ensure that she's in peak form when she encounters a vampire or demon. Small wonder, then, that the martial arts constitute a major component of every single episode in both the *Buffy* and *Angel* series. And all martial arts, at heart, are about effectively exploiting basic physics principles.

FINDING THE CENTER

In "What's My Line?" (B-2), Buffy finds herself battling a series of assassins Spike has hired to kill her—or, at the very least, to keep her occupied so she can't foil his nefarious plans. She makes quick work of the first assassin, but then finds herself trading blows with a young girl as strong and agile as she is. It soon becomes apparent that the girl is not one of the assassins. Instead, she turns out to be a second Slayer, named Kendra, who has mistaken Buffy for a vampire.

Both Slayers instinctively drop into traditional fighting stances when Buffy calls a time-out to assess this unprecedented situation. Their feet are spaced roughly shoulder-width apart to form a solid base, with one foot slightly forward, and their respective bodies are more or less erect, fists chambered—resting, just like cocking the trigger of a gun, on each Slayer's hips in preparation for a strike should the other make a sudden aggressive move. Any well-trained Slayer could tell you that the key to a powerful martial-arts technique is a strong fighting stance that provides good control of her balance. It has to do with a person's center of mass (also known as the center of gravity), typically located about one inch below the navel. Practitioners of Japanese martial arts call this area the *hara*, and consider it the seat of a person's power.

From the perspective of physics, this isn't far off the mark. Where Buffy's center of mass happens to fall is directly related to how stable her body position is, which is in turn related to how much force she can generate with her techniques. She achieves the most stability when her center of mass is located vertically over the center of her base, because her weight is distributed evenly on both legs. This makes it much harder to unbalance her. If her feet are too close together, her base is significantly narrower, and Kendra can more easily knock her off balance. Similarly, if Buffy bends at the waist as she punches Kendra, she reduces her own stability and makes herself vulnerable to a counterstrike by essentially leading with her own face. The importance of body stability is even more obvious when Buffy fights Angelus—Angel's evil alter ego—while she has the flu in "Killed by Death" (B-2). The high fever throws her off her game, and she swings wildly while punching, throwing herself off balance sufficiently for Angelus to briefly gain the upper hand.

Balance and stability are just two of the physics principles underlying even the most basic martial arts techniques. Let's consider what's involved if Buffy wants to throw a simple straight punch at Kendra. First, she must overcome her fist's inertia by exerting some kind of force. According to Newton's first law of motion, an object in motion tends to stay in motion, while an object at rest tends to remain at rest, unless an outside force intervenes. Every object—Buffy's fist, Kendra's body—has a given mass, and this mass determines its inertia: how much force is required to set an object in motion. Various muscles in the body must work together to produce movement. Energy is stored as a muscle contracts, and that energy is then transferred to another muscle when the contraction is relaxed, and so on. When Buffy throws a straight jab at Kendra's face, she generates a force through a series of muscular contractions that causes her fist to accelerate. To phrase it another way, when Buffy chambers her fist, her muscles contract. She is storing potential energy, which is converted into the kinetic energy of motion as she relaxes those muscles to execute the punch.

How much force is generated depends in part upon Newton's second law of motion: Force equals mass times acceleration ($F = ma$). The more force you apply to an object, the greater the rate of acceleration. So the harder Buffy throws her punch, the more her fist will accelerate. This is where proper stance comes into play. It's less about her overall body mass, and more about how much of it she can involve in her strike. Buffy's center of mass is located in her hip and abdominal region, which accounts for roughly one-third of her body weight. So more of her body mass is behind her punch when Buffy is stabilized than when she bends at the waist. And more mass translates into more force to accelerate her punch.

Buffy can increase the force behind her punch even further by stepping in, or snapping her hips forward as she punches. She might stun or disorient Kendra with short jabs to the face, thus buying herself time while the other Slayer is off-guard. But she can actually move Kendra backward if she puts as much of her body mass as possible behind her strike. There are other ways Buffy can exploit her body mass. She can bend her knees and sink into a low stance, storing potential energy like a coiled spring, and then quickly convert that into kinetic energy by thrusting her body upward to strike her target. Or she can lift her weight and then drop her entire body onto her doubled-over opponent to deliver the blow, getting a little extra help from the earth's gravity to achieve a greater amount of force. In fact, at one point in her exchange with Kendra, Buffy leaps onto a nearby table before executing a kick, gaining just enough extra height (and potential energy) to put a bit more "oomph" behind her kick as she jumps off.

Of course, Buffy also has an extra dose of mystical energy at her disposal: her Slayer strength. Since energy and mass are equivalent, as we saw in the previous chapter, she can generate much more force than she would otherwise be able to generate using her mass alone. This comes in handy against weaker human opponents, but it won't help her as much against Kendra, who has the same infusion of Slayer strength, or against the myriad of supernaturally strong demons she encounters on her nocturnal patrols. It just gives her a fighting chance.

The second element in Newton's equation is acceleration. It's a tricky concept, best described as how much velocity (an object's speed and direction) changes over time. The mass of Buffy's fist and its velocity combine to produce momentum. The more momentum that builds up as her fist accelerates, the faster its final velocity, and the more energy she can transfer into Kendra when she strikes. When Buffy's fist is chambered, its velocity is zero. As she executes her punch, her fist accelerates, gaining momentum as its potential energy is converted into kinetic energy. Since kinetic energy increases with the velocity squared, this means that if Buffy's fist is traveling twice as fast by the time she hits her target, it will have four times the kinetic energy, and she can transfer that much more energy into Kendra's body.

How long it takes to transfer momentum is also critical. Newton's third law says that momentum is conserved: It can neither be created from nothing, nor destroyed, but is passed from one object (Buffy's fist) to another (Kendra's body). We can combine

E=1/2mv²
F=ma

this with the concept of work, which can be calculated by multiplying the amount of force generated by the distance Buffy's fist must travel—and the time it takes to do so. There is a fixed amount of energy at Buffy's disposal, so she has to make a choice: Either she can transfer a small amount of force continuously over a longer period of time, or she can transfer a large amount of force in the shortest possible time. Either way, the total amount of energy delivered to Kendra's body remains the same, but unless Buffy plans a slow, prolonged siege against Kendra, she will try to transfer as much force as possible, in as short a time as possible, by executing her punch as quickly as she can.

The opposite applies to how Buffy lands when she somersaults over Kendra to avoid being struck by an ax. Here, the objective is to lengthen the time of impact. If Buffy locks her knees as she lands, her momentum drops to zero suddenly, and she feels a large force in her legs, possibly more than the bones and joints can handle. But if she lands with her knees bent and then rolls forward, she prolongs the time in which the impact takes place, so her momentum decreases more gradually, and the force of impact is smaller.

These same basic concepts apply to kicks. When Buffy throws a front kick at Kendra, she propels her body forward by pushing off her rear leg and bringing her forward leg into chamber position. The leg muscles contract and store potential energy. That potential energy is then converted into kinetic energy as Buffy executes the kick. The muscles contract and then relax, transferring the stored energy to the next muscle required to perform the kick. Buffy can generate even more force by pushing her hips forward, using her center of gravity to put more of her body mass behind the kick. It also slightly increases the distance her foot must travel before it hits the target—a seemingly small detail that can nonetheless translate into a stronger kick, since her foot accelerates more, and hence deposits more energy into her target.

BOARDS DO HIT BACK

What is happening at the point of impact, from the perspective of the target? If momentum is conserved in Newton's third law, then for every action there is an equal and opposite reaction. That is, if one object exerts a force on another for a given amount of time, the second object reacts by exerting an equal but opposite force for the same amount of time. The force generated by Buffy's punch creates a reaction force in the opposite direction when she hits Kendra. Kendra's body gains exactly the amount of momentum that Buffy's fist loses, barring any that is lost through conversion into other kinds of energy, like heat or noise.

Without a strong stance, Buffy's body might be pushed backward by the impact, making her punch much less effective. That's because part of the total energy is diverted from the target. If more of that energy is diverted than is deposited into Kendra's body—or if Kendra adds her own countering energy by stepping in to block Buffy's punch—Buffy may actually be forced backward. In fact, in "Revelations" (B-3), Buffy hurls herself into a double-leg flying kick at the demon Lagos. Lagos simply has too much body mass, and hence reflects the momentum in Buffy's kick, so she bounces right off him, to comical effect.

Newton's law of equal and opposite reaction applies not just to human or demonic bodies, but to every object—even the ground beneath our feet. Kung fu master Bruce Lee once famously observed, "Boards don't hit back." But in the strictest physics sense, they do. When a karate practitioner breaks a wooden board with his fist, he transfers momentum to the board, which accelerates in the opposite direction in response. If the part of the board that is hit—usually the center—is infused with more energy than its structure can handle, it will crack or break. A similar effect can be seen in "Smashed" (B-6), during an extended fight sequence between Buffy and a semireformed Spike, who have been inching toward consummating their long-standing love/hate relationship. The two former enemies trade blows and sarcastic taunts, tossing each other through the decrepit doors and into the sagging walls of an abandoned building nearby. The scene culminates in a spot of violent lovemaking that literally brings the house down around them.

Why doesn't Spike simply bounce off when Buffy hurls him against a wall? After all, that's what happens when Spike tosses *her* into a wall during their ferocious foreplay. But when Spike hits, the wall cracks instead. This has to do in part with their difference in mass. Just like Buffy's fist executing a punch, when a body is in motion, it builds up a certain amount of momentum based on the body's mass and velocity. Since Spike has more mass than Buffy, he gains more momentum, and thus more energy is transferred to the wall when he hits it.

Let's assume that Spike weighs 140 pounds and is traveling at a final velocity of 10 mph when he hits the wall. At that speed, his body would have about 504 joules of energy, equivalent to how much energy it would have if Spike just ran at the wall as fast as he could. That's actually not much energy, and were the wall made of brick, it's more likely that Spike would break a bone or two, rather than cracking the wall. In contrast, a two-ton car traveling at 10 mph upon impact would easily crack a brick wall. Fortunately the wall in question seems to be made of wood and plaster, and is in serious disrepair.

But why, exactly, does the wall crack? The wall accelerates in response to the added momentum from Spike's body, producing an equal and opposite reaction. But it doesn't accelerate uniformly. The areas that took the brunt of Spike's body accelerate much more than the surrounding wall, producing a localized strain. The fact that the walls are held in place by joints, foundation, and various supporting structures increases the strain, which eventually becomes too great. The effects are cumulative. The repeated infusions of energy brought about by Buffy and Spike's amorous thrashings create so much strain on the entire building structure—already in an advanced state of decay—that it begins to collapse around them. Rafters and debris fall to the floor, cracking the floorboards sufficiently so that when Buffy and Spike fall backward in an embrace, they crash through and fall into the basement.

What about objects made of something other than wood and plaster? Different materials can withstand different amounts of deformation, a property known as elasticity. Most materials are elastic to some degree: when they are deformed or bent by an infusion of incoming energy, they will bounce back to their original shape. But elastic materials all have their limits. Metal springs and rubber bands are very elastic. Plaster and glass are not very elastic; instead, they are brittle, and snap with even a small deformation. Energy, like momentum, is conserved, but in a collision, it can turn into different forms of energy, such as heat or noise. How much of the energy is converted depends in part on both the relative toughness and elasticity of the materials involved in the impact. There is no such thing as a perfectly elastic collision, but if there were, all of the energy would be transferred to the target with nothing lost to heat or noise, for example.

The *über*vamp (Turok-han) sent by the First to take out the remaining Potential Slayers in "Bring On the Night" (B-7) has an unusual natural body armor protecting its heart, on a par with a Kevlar vest, rather than the flesh-and-bone breastplate of the average vampire. That body armor is both extremely tough and highly elastic. When Buffy drives a stake into the Turok-han's chest, it becomes embedded, even though it clearly hasn't pierced through to the creature's heart. When the creature pulls out the stake, the body armor repairs itself. It rebounds from the impact and flows back into place, with no permanent rip or even an indentation where the stake hit. The same thing happens when Buffy shoots it with a crossbow in "Showtime" (B-7): The stake embeds itself, and the underlying breastplate is not pierced.

Anya compares the body armor to steel in "Empty Places" (B-7), but while the breastplate might share some properties with steel, the substance covering it has more of a doughy texture. The *über*vamp is like a very tough Pillsbury Doughboy or, to borrow Buffy's labored metaphor for herself in "Chosen" (B-7), cookie dough that isn't yet fully baked. Such a substance is considered "viscoelastic," making it very difficult to pierce. It easily absorbs the impact of the incoming stake, flowing around the intruding object so that the stake becomes embedded before it can penetrate the underlying breastplate and do any fatal damage. In fact, it appears to be a subset of viscoelastic material known as an "anelastic" solid: The substance recovers its former shape fully after the offending stake is removed.

Back in the real world, scientists at the University of Delaware have adopted an equally novel approach to improving Kevlar vests. They treat the fabric with a "shear-thickening fluid"—basically a syrupy, viscous mixture (similar to hagfish slime) consisting of silica particles suspended in polyethylene glycol—making it strong enough to stop a bullet, yet flexible enough to wear comfortably. This turns the fabric into a kind of "smart material," a class of materials that can sense and respond to changes in the environment—usually the presence of electric or magnetic fields, or changes in temperature.

In this case, the material responds to a change in pressure resulting from mechanical force. Under normal conditions, its molecules are weakly bonded and can move around with ease; that's why the material is so flexible. But the shock of an impact causes those chemical bonds to strengthen so the molecules lock into place; once the force from the impact dissipates, the bonds weaken again. That's why the fabric becomes rigid instantly when a bullet strikes, thereby preventing that bullet from penetrating, and reverts to its more flexible state once that force has ceased. A British company called d3o Labs manufactures "smart armor" using the same kind of material; it was used in the 2006 Winter Olympics to protect U.S. and Canadian skiers from injury.

Despite the ingenious design of its body armor, an *über*vamp can be staked if one stabs the region in just the right way so that the viscoelastic surface ruptures, enabling the wooden stake to pass through the breastplate to the heart—assuming that the stake doesn't splinter upon impact with the steely breastplate. Buffy simply needs to transfer much more mechanical energy than usual in order to penetrate both protective layers. Physics can help her a little in this instance, because the striking target is so small and focused. The same amount of energy focused into a smaller area (just over the heart) delivers more energy per unit area. A larger target surface (the *über*vamp's entire chest) would disperse the energy over the whole area and weaken the force of Buffy's strike.

Buffy can increase her chances of penetration even further by choosing to whittle her stakes out of the right kind of wood. Any woodworker could tell you that certain woods are harder, and split less easily than others. Pine, for example, is quite soft, with broad grains, which is why pine boards are used so frequently in martial arts breaking demonstrations. Maple and ash are denser, with very fine grains, and don't break very easily, which is why most *bokken* (long staffs used as weapons) in Japan are made out of ash. A stake carved out of ash would be less likely to splinter on impact with the bony breastplate of an ordinary vampire. Fortunately for the Scoobies' stake-whittling needs, ash is among the species of trees common to the Southern California region, so the wood should be in plentiful supply near Sunnydale.

There's another, less-well-known property of ash that makes it an excellent choice for staking the much-tougher *über*vamps. Since the 1880s, scientists have known that quartz crystals will produce a tiny voltage (an electrical field) when squeezed or pressed. They are another example of "smart" materials. This property is known as piezoelectricity, and it is an integral component of many modern technologies from sonar, radio, and television to the electric cigarette lighters found in cars and the portable sparkers used to light gas grills and stoves. The piezoelectric process also works in reverse. Applying an electrical field to quartz crystals will cause the crystals to deform ever so slightly, about one-billionth of an inch.

Piezoelectricity is not limited to crystals, either. While they would technically not be classified as smart materials, rubber, wool, hair, silk, bone, and certain kinds of wood nonetheless all exhibit (to a lesser extent) some piezoelectric properties: mechanical force, or stress, will produce varying amounts of voltage. It just so happens that ash is among those types of wood. A stake made of ash could conceivably spark-cut a hole upon impact, creating a breach in the Turok-han's tough-yet-doughy body armor and allowing the stake to pierce the steely breastplate and hit the *über*vamp's heart.

This might explain one of those pesky inconsistencies one encounters occasionally in the Buffyverse. The first time Buffy tries to stake an *über*vamp, she fails to penetrate the breastplate sufficiently to pierce the creature's heart, yet by the series finale, every one of her newly activated Slayers is able to reduce *über*vamps to dust with a stake (although these are not the only weapons at their disposal). Buffy could have used a softer birch stake—which has a much smaller piezoelectric effect, and hence produces a smaller voltage—in the first instance, then switched to ash for the final battle against the First's vicious horde, in hopes that ash's stronger structure and innate piezoelectric properties could even the odds a bit.

PUTTING A SPIN ON IT

"What's My Line?" (B-2) concludes with Buffy and Kendra joining forces against Spike and his minions, battling it out in an abandoned church. Making good use of the sacred objects at hand, Buffy takes out Spike by swinging a heavy metal incense burner on a chain (called a censer) around her head, like a lasso, then releasing it into the air. The censer hits the back of Spike's head with sufficient force to knock him out cold. This is a classic example of the physics of circular motion. It occurs when an object like the incense burner rotates around a vertical axis at the circle's center, such as Buffy's body.

It's a common misconception that what's at work here is centrifugal force, but in reality such a force doesn't exist. What keeps the chain taut as Buffy swings it overhead is the mass of the incense burner at the end of it, combined with the chain pulling the burner toward the center of the circle. This is known as a centripetal force. Without the chain pulling it inward to counter its natural inertia, the burner would simply fly through the air until gravity and friction from air resistance—or a collision with Spike's bleached-blond head—caused it to fall to the ground.

The principles behind circular motion are similar to those for linear motion. First, there is rotational inertia: an object's mass determines how much force is required to get it moving in a circle, or to change the motion of an object that is already rotating. Then there is an object's angular momentum, which is determined by multiplying how much force is required to change an object's motion (rotational inertia) by the rate at which it turns (angular velocity), for example, the number of revolutions per second. It is analogous to how mass and velocity combine to produce linear momentum.

Let's apply these concepts to the battle in the abandoned church. When Buffy begins whirling the incense burner around her head, she is applying a force to set the censer in motion. The now-rotating censer has angular momentum. Angular momentum is conserved—that is, remains constant—unless an outside force acts on the rotating object. In this instance, the censer's angular momentum increases because Buffy's effort is adding energy to its rotation, causing it to spin faster and faster. All that pent-up energy is released quite suddenly as she tosses the censer at Spike. As with linear motion, the more it accelerates, the greater the final angular velocity of the censer when Buffy releases it, and the more angular momentum it possesses. The resulting energy is sufficient to send the censer flying through the air with enough force to knock Spike out, since a large part of that energy is transferred to his head.

Before she comes up with this ingenious method of taking out her opponent, Buffy throws a good number of punches and kicks at Spike, including a spinning back kick. Most spinning techniques in the martial arts rely on the same basic principles of circular motion. First, Buffy must generate sufficient force to overcome her leg's inertia. The force that is applied to a rigid object, like Buffy's leg, is called torque. It is simply a rotational (twisting) action, such as the force one would apply to tighten a bolt with a wrench. Buffy must press against the ground to start the rotation for the kick, pushing off hard to exert enough torque to propel her body around. In doing so, she transfers energy to the ground, which pushes back in response. She can generate even more initial torque by twisting her body in the opposite direction before throwing the kick. She literally "winds up" to produce more torque, much like an ice skater does before executing a jump. (On the downside, this will "telegraph" her intentions to her opponent, robbing her of the element of surprise.)

Once Buffy's body is in motion, her angular momentum remains constant. Whatever energy she has managed to generate with her torque can't be increased any further. But there is a way to increase the angular velocity (spin rate) of her kick. Unlike plain old inertia, Buffy can change the speed of a spinning kick, simply by changing how her mass is distributed. The farther her foot is from the center of the pivot point—for instance, if her leg is fully extended horizontally as she kicks—the more rotational inertia it has, and the more force is required to start or keep it moving. The closer to the center, the less rotational inertia her foot possesses. This also applies to the spinning incense burner: Buffy can increase the rate of its spin simply by choking up on the chain to make the circle smaller. The censer's mass is now closer to the circle's center than before, and therefore spins faster.

That's why Buffy "coils up" as she spins into the kick, bringing her arms tight into her body and chambering her leg—again, just like an ice skater executing a jump. Buffy only opens up when she strikes. A closed position, with the arms and legs pulled in tight against the body, decreases her angular inertia and increases rotation speed, while her angular momentum remains constant. So Buffy's angular velocity must increase to balance things out: She spins faster to make up for the lessening of rotational inertia. On the other hand, an open position, in which the arms and legs are allowed to swing away from the body, causes the speed of rotation to decrease. All that energy of motion is released when Buffy uncoils and throws out her leg for the actual kick. Because her foot has more distance and time to accelerate than with a linear front kick, it reaches a greater final speed at the moment of impact. And once again, the greater the final velocity, the more energy Buffy can transfer into Spike's body when she kicks.

BALANCING ACT

As welcome as the sight of a high school self-defense class might be in "Phases" (B-2), it's distressing that the gym instructor fails to correctly teach a basic hip throw. In fact, she inadvertently provides a useful lesson in how *not* to throw one's opponent. When the school's resident leering would-be Lothario, Larry, grabs Buffy from behind to simulate an attack, the gym teacher intones, "Bend over, using your back and shoulders to flip your assailant to the ground." Not surprisingly, only Buffy is able to do so, flipping her much larger attacker over to the ground quite easily, thanks to her Slayer strength—plus a little extra incentive to violence when Larry lasciviously grabs her derriere.

What's so bad about using your back and shoulders when executing a hip throw? Nothing, if you've got Larry's linebacker build, or Buffy's Slayer strength. Most high school girls lack these crucial attributes, however, and would probably hurt themselves if they tried to throw an attacker the way the gym teacher taught the technique. But with the aid of basic physics, any one of the teenage girls in the class can use an attacker's size and strength against him to toss him to the ground.

$E_r = \Omega LM$

Buffy's gym teacher clearly didn't understand that just like punches and kicks, the true power behind any hip throw comes not from the back and shoulders, but from a person's center of mass: the point where gravity acts on the body as a whole. Think of the seesaw in Sunnydale's playground. It has a triangular base that acts as a fulcrum, while the long board lying across it acts as a lever, pivoting back and forth above the fulcrum. The point at which the board is perfectly balanced atop the fulcrum is its center of mass: The lever is divided into two equal halves. Move that point ever so slightly in either direction, and one half of the lever becomes greater than the other half. The board will tip out of balance.

Balance is the essence of judo. Larry is balanced, and therefore stable, as long as his center of mass remains over his feet. If Buffy moves Larry's center of mass outside of his stable base, gravity will make him lean or fall, and hence he will be easier to throw. This is where torque once again comes into play. A key element in determining torque is the length of the lever: the perpendicular distance from the pivot point (Buffy's hip, Larry's center of mass) to where the force is being applied. Once Buffy breaks her opponent's balance forward, even if she does nothing more, gravity will pull on his center of mass. Since his center is now unstable, this gravitational pull creates a torque, causing rotational motion, and he can topple over all by himself. The more he is leaning forward, the longer a lever his body forms, and the greater the resulting torque. When Larry is standing upright, for example, his center of mass is stable, and his torque is therefore zero. If his center of mass is off balance, the resulting torque causes his body to rotate. Buffy can then reinforce this rotation by pulling on his arm to apply even more torque, while using her hip—placed just under Larry's center of mass—to lift his feet off the ground.

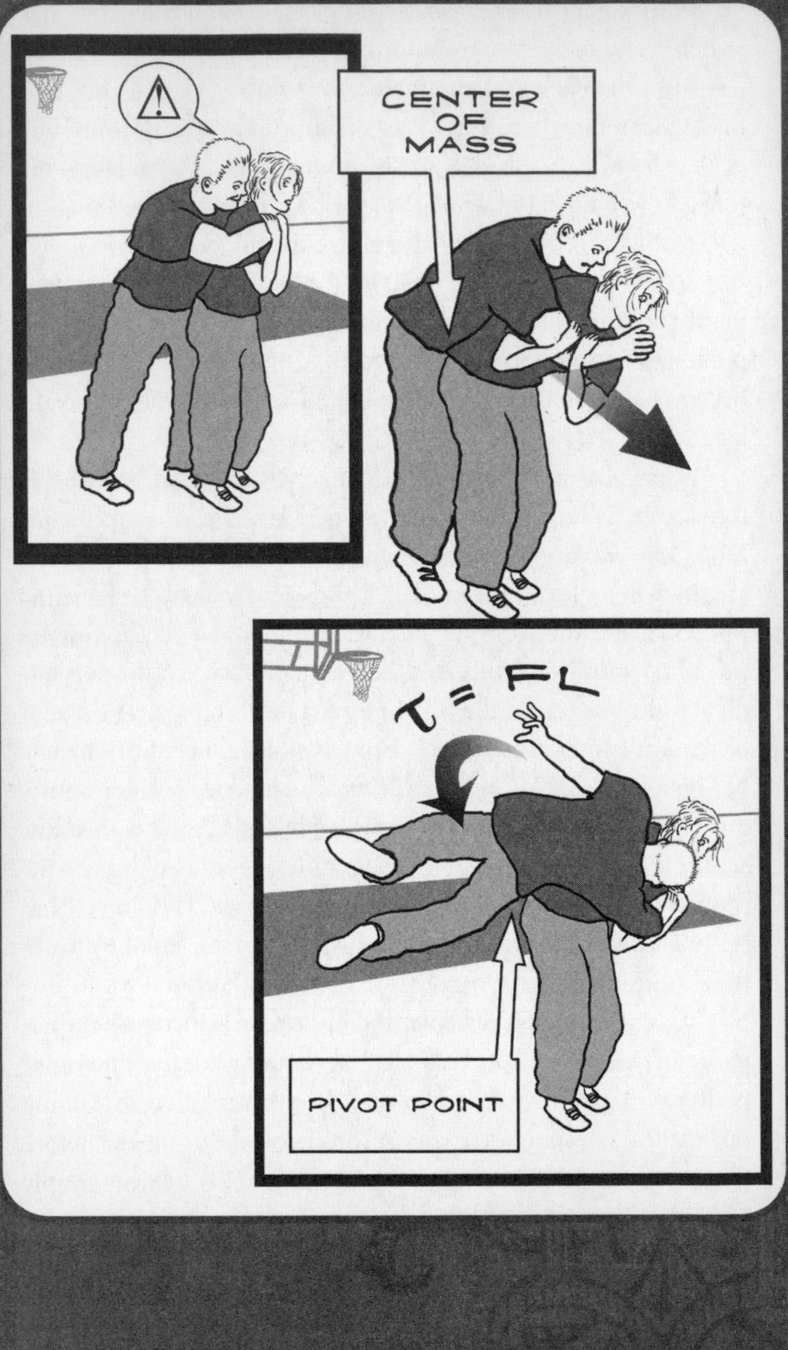
CENTER OF MASS
τ = FL
PIVOT POINT

The concept works even better if Larry is walking forward. Any kind of movement requires the body to be off-balance at some point. As Larry walks, his foot comes off the ground as he steps forward, and for that moment, he is off-balance—technically, he's falling. If Buffy turns her body in toward Larry quickly as he walks, and fits her own hip just under his center of mass, she can redirect the energy from his forward movement, easily lifting his feet off the ground, then pulling him over and around, using her hip as the pivot point around which Larry's body, as the lever, will rotate. The faster Larry is moving toward her, the more forward momentum he gains, and the easier it is for Buffy to throw him. In essence, she uses his own energy against him.

When Buffy initially tries to throw Larry in her self-defense class, he is stationary, so she can't take advantage of any forward momentum he might generate by walking toward her. And his center of mass is not leaning out far enough for him to be sufficiently off-balance. His body mass is large enough to oppose the torque she is trying to apply by pulling with her back and shoulders. Buffy overcomes this through sheer Slayer strength. That might work for her against a human—powerfully built though he is, Larry's strength is limited by his mortality—but she is unlikely to be able to throw Angel or even the lithesome Spike in this way. They have more body mass than she does, plus the same extra infusion of mystical strength to put behind it. The larger and stronger her opponent, the more force will be required to move him. That's why good technique is so critical.

The good news for the diminutive Slayer is that hip throws generally work best against a taller opponent like Angel, because it's much easier for her to get under his center of mass—assuming that she can catch him off-balance and execute the throw before he has time to counter it. (Angel can counter the hip throw simply by bending his knees to lower his center of mass, "planting" his feet into a solid wide base, and leaning slightly back to counter Buffy's forward pull.) Bending her knees slightly as she turns her hip in toward Angel stores potential energy, which is then converted into kinetic energy when she straightens them. This extra burst of energy is just enough to set Angel's body in motion for the throw. Once again, mass and velocity are the critical factors in determining how hard Angel falls. The faster Buffy throws Angel, or the heavier he is, the greater the force of impact when he hits the ground.

As with striking techniques, proper stance is critical for throwing. Buffy should stand with her feet shoulder width apart so that she is stable when she begins to execute the throw. But once the throw is in progress, her feet should be as close together as possible, heels together and toes pointing outward to form a V. She should become a human fulcrum. It might feel stronger to be standing in a wider stance—and indeed it is, if Buffy's sole purpose is to maintain her own balance—but once the 200-pound Larry is hoisted onto her back, any space between Buffy's legs provides a direct path to the ground via gravity's pull. Her knees could easily buckle under the extra weight.

Buffy can always choose to let gravity do *all* of the work for her in a fight. This is how Illyria deals with Angel in "Shells" (A-5). As Angel lunges for her, Illyria sidesteps nimbly out of his line of attack, grabbing his jacket and continuing his forward momentum by tossing him through a plate-glass window. It's not an actual throw, per se, but she does employ the judo principle of using her opponent's momentum against him, and basic physics does the rest. Once through the window, Angel falls several stories to the pavement below. The higher the floor of Wolfram & Hart's headquarters on which Angel is standing, the more potential energy he starts out with, the more momentum he gains as he falls, and the greater the resulting kinetic energy when he hits the ground—because he is falling over a greater distance, and for more time, than if he simply fell from the second floor.

All bodies fall at the same rate regardless of their relative mass (barring the presence of air friction), so Angel would speed up by 9.8 meters per second for each second that he falls. If Angel falls for five seconds, he will reach a speed of about 50 meters per second. He doesn't fall quite that far, but even falling several stories, he would be traveling at a speed of 30 meters per second. When he hits the pavement below, the ground will respond with a counter force equal to the force Angel's body exerted on it at the moment of impact. The force will transfer the energy of Angel's fall into the pavement, minus any energy lost to heat and noise. If the energy transfer is large enough, Angel's body might even crack the concrete pavement.

In this case, the energy would be about 40,500 joules, the same amount of energy contained in 1/50th of a stick of dynamite. That might not sound like much, but any human body falling that great a distance would be literally shattered by the impact. We can only assume that Angel's vampiric powers allow him to survive the fall—that, and the fact that he lands in a forward break-fall position. Instead of one part of his body hitting first and bearing the full brunt of impact, his entire body lands horizontally, with his arms out to the side, palms down, and his face turned to one side. This disperses the impact over a wider surface area (and ensures that his nose isn't crushed). Angel will still suffer broken bones, not to mention internal injuries, but these are trivial to an immortal vampire with a mystically rapid healing rate.

Hip throws aren't the only weapon in judo's impressive arsenal of techniques that have wormed their way into the Buffyverse. In "Lover's Walk" (B-2), Spike returns to Sunnydale, only to be waylaid by a "committee" of vampire goons hired by Mayor Wilkins. An all-out brawl ensues. Amid the usual flurry of flying fists and feet, Spike deftly "clotheslines" an attacking vampire across the throat with his arm, throwing his opponent backward to the ground. As the vampire goon attacks, his entire body is moving forward at roughly the same rate. When Spike's arm hits the goon's throat, he exerts a force on the vampire that is far from its center of gravity. Specifically, Spike's arm exerts a torque on the vampire that converts some of its forward motion into rotational motion. The goon's feet fly out from underneath him while gravity does the rest, pulling his top half to the ground.

It's similar to what happens to a seesaw if its pivot point is moved way off to one side. The seesaw is no longer balanced between equal halves, and the heavier end slams into the ground. Because all the energy comes from the goon's forward momentum, Spike exerts almost none of his own energy. He generates just enough force to redirect the energy of his opponent's attack. The more forcefully the other vampire hurls himself at Spike, the faster he accelerates, the more quickly he topples over, and the harder he hits the ground—and the less of Spike's own energy is required to execute the technique.

SIZE MATTERS . . . SORT OF

The testosterone-ridden werewolf hunter who taunts Buffy in "Phases" (B-2) isn't alone in his skepticism when it comes to a girl's ability to fight. There's an oft-repeated phrase in martial arts circles: "All things being equal, the bigger, stronger person will win." That statement has some scientific merit. Someone who is heavier, like Angel, will be able to generate the same amount of force as a smaller person, like Buffy, using much less effort, even though both possess roughly equal amounts of technical proficiency and extra mystical strength.

The more force that acts on an object, the faster it will accelerate, yet the more massive an object is, the greater its inertia, and the more it will resist acceleration. If Buffy throws a punch at Angel, it won't have as much impact as that same punch would have on the much smaller Kendra, because how much he can accelerate is inversely proportional to his mass. That is, it takes more energy to move him than it does to move the other Slayer. He has more mass than Kendra, and hence more inertia. Buffy can produce only a fixed amount of energy. She must transfer more of that energy into Angel's body to move him than she would have to use against Kendra.

The only way to increase the force of Buffy's punch is by increasing either her mass or the rate of acceleration. The average martial arts technique is executed in mere seconds. It's physically impossible for Buffy to consume enough calories, or pump enough iron, to gain significant body mass over such a short period of time. So Buffy's mass will be more or less constant, although she can gain some power by throwing more of her total body mass behind the technique.

Making matters worse, the faster an object accelerates, the more energy is required for it to keep accelerating, until it reaches a point where it can accelerate no more. Thus, although good technique can help increase the speed of Buffy's punch, it can only do so to a certain degree. Sooner or later, she will reach the upper limit of a technique's potential for acceleration, and she may still fall short of the final velocity necessary to overcome the substantial difference in mass between herself and Angel. Physics is innately unfair in this respect. Even in judo, which was designed to help the weak overcome the strong, size still matters when the competitors are equally skilled. Modern judo competitions are divided into weight classes for that very reason.

If that's the case, why doesn't Buffy simply bow to the inevitable when faced with an opponent who possesses superior strength: Glory, the exiled hell god, for example, or the misogynistic preacher-turned-serial-killer, Caleb, who derives his superhuman strength directly from the First? Anyone can do the math and determine that Buffy is destined to lose. Yet again and again, she prevails. When Buffy faces off against a former high school classmate-turned-vampire in "Conversations with Dead People" (B-7), he challenges her confident assertion that she will win their little death-match: "Two years of tae kwon do plus vampire strength—I think somebody's counting their chickens." While he is bigger and stronger, and has some formal training, Buffy has more practical combat experience, and hence knows something he doesn't: All things are *never* equal. Size matters, but it's not the whole story.

A popular physics joke describes how a farmer, a psychologist, and a physicist are asked for their advice on how to cajole a dairy cow into giving more milk. The farmer suggests revising the animal's diet. The psychologist suggests painting the cow's stall in soothing colors. And the physicist's approach begins with the statement, "Assume the cow is a sphere . . ." Physicists will often make these kinds of simplifying assumptions. It makes their calculations a bit easier, and the results are sufficiently accurate for their limited scientific purposes. Nonetheless, it is an idealized construct. In the real world, a cow will never be a perfect sphere. And a street fight will never be fought under the same carefully controlled conditions as in a martial arts competition. There are no rules, no unspoken code of honor, and no mercy. In the Buffyverse, the combatants are truly trying to kill one another.

The Buffyverse, like the real world, is not a controlled environment. It is nonlinear and highly chaotic, and thus there are any number of other variables that can determine the outcome of a fight. For example, deprived temporarily of her Slayer powers on her eighteenth birthday ("Helpless," B-3) in a sadistic coming-of-age acid test known as the *Tento di Cruciamentum*, Buffy must defeat an especially vicious vampire by relying on her wits, not her strength. (She tricks him into drinking holy water.) After being soundly beaten by Caleb the first time they trade blows ("Dirty Girls," B-7), she switches tactics, using her speed and agility to dodge and roll until she gets hold of a mystical scythe that proves to be the equalizer between them ("End of Days," B-7). She knows how to exploit the element of surprise, using the Buffybot as a decoy against Glory to catch the hell god unawares ("The Gift," B-5). And she is resourceful in a pinch. Spike notes as much when studying a videotape of her battle with a nameless vampire in "Halloween" (B-2). Lacking a stake, she improvises, grabbing a nearby wooden fence post to reduce the vamp to dust.

The downside of all this unpredictability is that a single moment of hesitation, the smallest mistake, can cost the Slayer her life—and, sooner or later, the law of averages dictates that every Slayer will make that fatal error. As Spike fights an unnamed Slayer in turn-of-the-century China during the Boxer Rebellion in "Fool for Love" (B-5), the Slayer loses focus for one split second, right when she is about to stake him. That's all it takes for him to gain the upper hand and sink his teeth into her neck. In fact, a Slayer's margin for error decreases in inverse proportion to the difference in size and strength between her and her opponent. She will have less room for error against a vampire who outweighs her by 60 pounds than against one who outweighs her by 20 pounds. And sometimes a fight can be determined by something as intangible and unscientific as the relative mind-sets of the combatants. "It's not about the moves, love," Spike tells Buffy when she asks him to describe how he killed two Slayers. "Every Slayer has a death wish. Even you."

Fortunately, Buffy's alleged death wish is offset by her many positive attributes. She is smart, creative, willing to buck tradition and risk trying something daring and new instead of doing things the way they've always been done—an aspect of her personality that repeatedly brings her into conflict with the stodgy Watchers' Council. She understands that it's not always wise to play by the Council's antiquated (and rather chauvinistic) rules, especially against demons who don't adhere to human codes of conduct. She has the support of friends and family. And while she struggles with self-doubt as much as anybody, she doesn't give in to defeatist thinking, even when she knows the odds are against her and she is likely to lose. Physics notwithstanding, that's the hallmark of a true champion.

5

TIME GOES WONKY

Relativistic Tricks of the Temporal Trade

"What is time? If no one asks me, I know.
If I wish to explain it to one that asketh, I know not."
—AUGUSTINE OF HIPPO

Time must seem so irrelevant to an eternal being. In "I Will Remember You" (A-1), Angel consults the Oracle, a brother/sister tag team that exists in a higher dimensional space outside the realm of time. He offers his watch as a gift. The Sister Oracle is delighted with her exotic new trinket, even though it is meaningless in her dimension, and Angel is granted an audience. He returns to his own world and finds that he has only been absent for a split second, despite his extensive conversation with the Oracle. Since the Oracle's realm lies outside of temporal reality, no time at all has passed beyond its borders.

This is just one example of the kinds of temporal anomalies that occur in the Buffyverse even though, for the most part, it observes a reasonably consistent chronological time. For instance, when either series breaks for the summer, a corresponding three-month interval is usually assumed to have lapsed when episodes resume in the fall. Halloween and Christmas correspond more or less to our real-world holiday seasons, and Buffy's birthday always falls in mid-January: she's a Capricorn on the cusp of Aquarius ("Doomed," B-4).

But occasionally, time goes wonky, wreaking all manner of havoc. Not only does the dimensional space of the Oracle lie outside of time, but there are also other dimensional worlds where time moves faster or slower than in Sunnydale or Los Angeles. There are episodes where the chronology of time runs amok, with events that seem to happen out of sequence instead of in succession, and time loops, where the same sequence of events occurs over and over again. There are cases where time folds back on itself. And there are creatures capable of warping time for their own nefarious purposes. While the Buffyverse is undeniably fictional, these anomalies can illustrate some intriguing concepts in real-world physics.

TIME'S ARROW

By its very nature, time is an intangible, elusive quality. Buffy learns this firsthand in "Life Serial" (B-6). Each member of the Trio devises a test to gather data on the Slayer's abilities and hopefully identify weaknesses. One of those tests causes temporal anomalies. It involves placing a small "inhibitor" device similar to a microscopic "bug" onto Buffy's sweater while she's standing in the hallway between classes at UC-Sunnydale. It supposedly emits a series of electromagnetic pulses that cause a localized "temporal shift" around Buffy. Suddenly everyone around her is moving at a much faster pace; time speeds up for everyone except her. When the bug self-destructs, the flow of time returns to normal. This is obviously not something we see in the real world. We go about our daily lives secure in the knowledge that time marches on, always moving forward at the same rate for everyone. But might this be an illusion? In theoretical physics, as in the Buffyverse, time can do some very strange things.

What exactly *is* this thing we call time? It's best to simply say that time is the measure of how things change. Around 360 BC, the Greek philosopher Plato wrote in his treatise *Timaeus* that time was born when a divine worksmith imposed form and order on primeval chaos. There is indeed something a bit arbitrary about how we measure time; after all, it's not like we can see it. The standard for time on which any measurement is based is chosen at random. Today, we set our clocks in accordance with Greenwich Mean Time, but that's just because in 1884 a group of scientists declared that site the prime meridian of the Earth. We can all unilaterally agree to change to a new standard whenever we like.

Equally arbitrary is the decision to mark time as 60 seconds to a minute, 60 minutes to an hour, and so forth. Historians aren't sure how it came about, but we can safely blame the Egyptians and Babylonians. Ancient Egyptians used a calendar with twelve 30-day months, yielding 360 days to the year. They may have gotten this notion from the Babylonians, who were very fond of the number 6 and used a base-60 numerical system. It seems an unlikely coincidence that dividing 360 by 6 equals 60, so it's likely that our units of time owe something to Babylonian mathematics. The notion of a second is generally believed to have once been equivalent to the period of about one heartbeat. Scientists like Galileo used to time experiments using their pulse—a method that would prove problematic for Angel, since the Undead have no pulse.

Time always seems to move forward, at least from our human perspective. Plato made a philosophical distinction between "Becoming" (the realm of time) and "Being" (an eternal, unchanging state). The Oracle, for example, exists in an eternal state of Being; meanwhile everything in the Buffyverse is Becoming. For Plato, time also apparently ran forward: We are constantly becoming, until we simply are, at which point there is no time. Other ancient cultures preferred to view time as a cycle of birth, death, and rebirth. Aristotle wrote of time moving in a circle, and the Mayans believed that history repeated itself every 260 years, a period they called the *lamat.* (The Mayans also believed that the universe would die and be reborn every five thousand years, in an eternal cycle of birth, death, and rebirth.) It wasn't until the Christian era that the idea of a clear, one-directional arrow for time was firmly established. Christ's death and resurrection were considered unique events, unrepeatable, and thereafter time only moved forward—at least in Western thought.

In terms of philosophy, the notion of time's arrow might have arisen out of Christian theology, but there is also a solid scientific basis for it. Thermodynamics is the branch of physics that deals with heat and motion, specifically, how heat is converted into other forms of energy. There are two fundamental laws in thermodynamics, as we saw in chapter 3. First, energy is conserved. Heat and work are pretty much interchangeable, and energy can be transformed, but can neither be created ex nihilo nor destroyed. It's the second law that directly affects our notion of time. The universe has a general tendency toward disorder and decay, a phenomenon known as entropy. Entropy dictates an arrow of time that runs only in one direction: forward. Heat only flows in one direction, so a cooler body can't pass heat to a hotter one any more than a melted ice cube can refreeze of its own accord. Time, therefore, is an irreversible process.

This should be a pretty straightforward conclusion, but mathematically, things are more complicated. Every major physics theory to date—Newtonian mechanics, special and general relativity, and quantum mechanics—exhibits something called "time-reversal symmetry" in its equations. In other words, the value for time (T) can be either positive or negative. The equations allow for the possibility that time can flow forward or backward, even if real-world physics precludes it because of entropy. The math says that there is no uniform direction in which time must always flow—a concept that is in direct opposition to our daily experience.

Yet we've also seen that entropy is a statistical, rather than absolute, law. In "Fool for Love" (B-5), Spike is shooting pool at the Bronze while regaling Buffy with the story of his Unlife. He makes the break at the start of the game, and the balls disperse across the table's surface. Using Newton's laws, he could easily determine where each ball would end up after his initial salvo, assuming that he isn't distracted by the arrival of his spicy buffalo wings. Similarly, if he could track the exact trajectories of each and every ball in reverse, and could then apply just the right amount of force with the pool cue, at exactly the right angle, every ball would end up right back where it started.

The inevitable energy loss from friction, the angle of the table's surface, and other external forces complicate matters in the real world, so Spike's feat is unlikely to occur, but it's not impossible. From a purely mathematical standpoint, there is a tiny statistical probability that Spike could play the entire game backward. He would merely have to re-create the motions of every single atom in reverse—an ability that would give the blond bloodsucker an almost divine level of omniscience. "For such an intellect, nothing could be uncertain, and the future just like the past would be present before his eyes," the eighteenth-century French mathematician Pierre Simon LaPlace once wrote.

Isaac Newton thought time was absolute, believing that at the creation of the universe God had wound up a "master clock," which has been ticking away ever since in a precise and predictable fashion. Time was an integral factor in his equations when he formulated his famed laws of motion in the seventeenth century. He defined motion as how an object changes position in time. A body in motion with no forces acting upon it will remain in motion, moving in a straight line at constant speed. According to Newton, that constant speed can be measured against an absolute time. His contemporary rival, Gottfried von Leibniz, believed in so-called relational time: Space and time merely described the relationships between where objects were and where events took place.

To get an idea of relational time, let's look at a scene from "Doppelgangland" (B-3), in which Oz's band, Dingoes Ate My Baby, is playing at the Bronze. Hours before the show began, both the stage and club were empty. Imagine a lone metronome ticking away in one corner of the empty stage; that is the preexisting absolute time that Newton believed governed the universe. Then the club filled and the band members came onstage and began to play. Now the Bronze is no longer empty. The Dingoes don't hear the metronome. They create their own kind of time. The drummer sets the beat, and the rest of the musicians follow his rhythm, as do the people dancing in the crowd. That is the "relational time" espoused by Leibniz.

The problem with Newton's concept of absolute time is that since it lies beyond human perception, there's no way we can ac-

$R_{mv} - \frac{1}{2} g_{mv} R = k T_{mv}$
ALMOST FINISHED.

curately measure it. Any number of timekeeping devices have been invented over several millennia, from ancient sundials and pendulum clocks to quartz clocks, but these different devices aren't even consistent with one another—at least not for long. The Scoobies can all synchronize their respective watches before their late-night mission to break into Mayor Wilkins's office in "Choices" (B-3)—or they would, if they'd remembered to wear watches—but eventually each will lose or gain time compared to the others. The atomic clock is the most accurate timekeeping device yet invented, since it relies on measuring the changing energy states of atoms, which serve much the same purpose as a pendulum, and are rarely influenced by external conditions. Yet even atomic clocks can be thrown out of synch.

Since absolute time is beyond our ken, we need a physics theory that works regardless of which kind of "clock" someone uses. A young physicist working in a Swiss patent office obliged in 1905 with a theory called special relativity. According to Albert Einstein, Newton was wrong (and Leibniz was on the right track). Time isn't absolute; it's relative.

RELATIVE TIME

We see a fairly accurate description of what happens to time under special relativity once Buffy completes the series of tests in "Life Serial" (B-6). The Trio compares scores to determine who designed the best test. Jonathan insists that he should be declared the winner, since his test of the Slayer—a time loop in which Buffy relives the same sequence of events over and over again—took the longest for her to complete. But Andrew points out that this is only from Buffy's perspective. From the perspective of Giles, who was unaware that the time loop was happening, Jonathan's test took the shortest time of all.

So time can slow down for one person but not for another. The technical term is "time dilation," and it's a direct consequence of one of the central tenets of special relativity: There is no such thing as a fixed frame of reference. In physics, a frame of reference simply denotes where a person or object happens to be standing, relative to the rest of the universe. There is no absolute frame of reference for time against which all motion can be measured. Einstein explained that this is because everything (and everyone) is constantly in motion through both space and time, and therefore everything has its own unique frame of reference. Two people who are moving relative to each other, wearing identical watches, will measure time differently. Time will slow down or speed up depending on how fast each is moving. Einstein's theory shatters Newton's concept of absolute time, in which events occur at the same time and point in space that is universal to everyone.

As weird as time dilation sounds, it does indeed happen. It just isn't noticeable in everyday life because standard watches and clocks aren't sensitive enough to measure the tiny discrepancies that appear at slower speeds. Time dilation only becomes significant at speeds approaching the speed of light, when its effects are greatly magnified.

Things get even weirder when we consider why this happens. Einstein asserted that space and time are one. Our three-dimensional existence—the "where" of an event—evolves along the fourth dimension of time—the "when" of an event—so we live in a four-dimensional space-time. What happens to time must therefore also happen to space. So as time dilates for an object in motion, the object's length contracts along its horizontal axis by a corresponding amount. For instance, both Buffy and

Spike become noticeably thinner as the series progresses; the actors playing them have clearly lost weight. Yet another (admittedly absurd) explanation is that they are moving faster each season, so that their horizontal length contracts, at least from the perspective of the television viewer. If Buffy and Spike were to approach the speed of light, they would become so thin as to appear almost two-dimensional to an outside observer, without giving up a single calorie. By the same token, it might be argued that the morbidly obese demon Balthazar, who appears in "Bad Girls" (B-3), isn't so much extremely fat as really, really slow.

Space and time might be in the eyes of the beholder, but the speed of light is constant, regardless of frame of reference. Since Newton, scientists had assumed that motion through space was completely separate from motion through time, but if space and time are one, then the two types of motion are inextricably linked. Time dilation and length contraction are the result of space and time adjusting with motion to ensure that two people moving relative to each other will always measure the same speed for light: 670 million miles per hour. It's helpful to think of it in terms of a tradeoff: As an object speeds up as it moves through space, time must necessarily pass more slowly.

Let's assume that the students at UC-Sunnydale really are moving at a much faster pace than Buffy when the "bug" is activated in "Life Serial" (B-6). A traveling clock ticks more slowly than a stationary one, and the faster a clock travels, the more slowly it ticks. So if Buffy were able to check her watch in the middle of her "temporal shift" and compare it to the time showing on the watches of the people racing around her, she would find that her watch runs more quickly than theirs. As those people speed up, a portion of their motion through time is diverted into motion through space, so time moves more slowly for them. Buffy, on the other hand, is mostly stationary.* Most of her motion is through time, and she experiences little time dilation. Bizarre as it seems, these effects ensure that no matter how fast Buffy and the students are moving relative to one another, everyone will measure the exact same speed for light.

So light is the link between space and time. It also sets the cosmic speed limit. Nothing can travel faster than the speed of light, because that is the point at which all motion through time is fully diverted into motion through space. The forward motion of time stops completely at exactly the speed of light. Furthermore, events can't be said to happen at the same moment by two observers in different frames of reference, because no one ever sees the world as it is right "now." We can't tell if something has happened until the information reaches us, and there is always a delay of at least the speed of light before that happens. How much it is delayed depends upon the relative speed of whoever is observing an event.

It's a difficult concept to visualize, but imagine that we're standing on the arrival platform at Sunnydale's train station in "Crush" (B-5) when the train pulls in. It is carrying a car full of dead passengers, thanks to the presence of the vampire Drusilla. When the train is stationary, it shares our reference frame. Suddenly, there is a flash of light at the center of the car—perhaps the result of the terrified conductor firing off a flare gun as an alarm upon discovering all the blood-drained bodies. The light emitted by the flare will reach passengers at either end of the car at the same time—a fact they would be able to verify, were they still breathing. Since we share the same frame of reference as the

*Even when standing still, Buffy stands on the Earth, and technically is "moving" with it as the Earth travels through space and time. None of us is truly stationary.

train, we, too, will see the light arrive at both ends of the car at the same time.

But what if the train just sped right past the station without stopping? (Who wants to stop in a town where the mortality rate is far above the national average, even with a resident Slayer?) Now we are in a different frame of reference from the dead passengers, who are moving relative to us because their reference frame (the train) is moving. Any (living) person sharing that moving reference frame would still perceive the flash of light from the conductor's flare as reaching both ends of the car at the same time. But to us, on the platform, the light will reach the passengers at the rear of the car before it hits those at the front. That's because the rearmost passengers are moving into the light beam, so it has less distance to travel, whereas those at the front are being carried away from the light beam, which has farther to travel to reach them. If two lucky surviving passengers at either end timed the light beam with their watches, those watches would verify that the light strikes each end of the coach at the same time. But that time would be different from the time recorded by the Sunnydale station clock on our platform.

Time dilation has another quirk: It allows for theoretical time travel into the future. That's the basis for Einstein's "twin paradox." We learn in "Damage" (A-5) that Buffy and Dawn are living in Rome after the final battle against the First. Say that Buffy jets off to China to collect a new Slayer, leaving Dawn at their Italian pied-à-terre. They synchronize their watches before Buffy leaves. When Buffy returns, she will compare her watch to Dawn's and find that time moved more slowly for her than for her sister. Technically, she returns to the future, even if it's just a few seconds ahead. Common sense might tell us that when Buffy and Dawn once again share the same frame of reference (their place in Rome), their watches should be in synch. However, Buffy and Dawn haven't had equivalent experiences. Buffy accelerates initially upon leaving Italy, then decelerates, turns around, and accelerates back for the return trip. Dawn inhabits an inertial frame of reference; her "speed" is constant (at least with the rotation of the earth). So the symmetry between them is broken.

This phenomenon has been experimentally verified. In the 1970s, scientists synchronized two atomic clocks and then sent one on a round trip on a jet airplane. When the traveling clock returned, more time had passed for the clock that remained behind than for the one that was aboard the jet plane. Of course, the difference was only a few seconds, and as Cambridge physicist Stephen Hawking has pointed out, "It would take an awful lot of frequent flyer miles to prolong one's life by a day this way." Still, the closer to the speed of light Buffy traveled, and the farther she went before turning around, the more time would slow down for her. Theoretically, she could return to find that Dawn is now the older sibling, instead of being five and a half years younger.

Unfortunately, special relativity doesn't allow for traveling *backward* in time. In order to travel into the past, Buffy would need to move faster than the speed of light—which is impossible. In fact, no object with mass can ever reach exactly the speed of light. As Buffy speeds up, her mass increases proportionately, so she would get heavier and heavier the faster she goes—although ironically, she would appear to be getting thinner and thinner, thanks to length contraction, from the viewpoint of an outside observer. Eventually her mass would become so great that it would take an infinite amount of energy to keep accelerating. But there is a theoretical loophole in this prohibition against visiting the past.

FOLDING BACK TIME

Angel consults the Oracle because he's been accidentally infected with the blood of a demon with the power to restore life, making him human again. But humans are fragile and mortal, and as such don't make the best champions in the fight against evil. (It's a miracle the humans in the Buffyverse last as long as they do without the benefit of mystical superpowers.) So Angel asks the Oracle to take back the day, restoring his vampiric nature. The Oracle does this by creating a temporal fold: Time doesn't so much run backward as fold back on itself, such that the last twenty-four hours never happened.

Conceptually, Angel owes his singular experience to Einstein, who extended his work on special relativity in 1915 to incorporate the effects of gravity and acceleration. Special relativity was primarily concerned with objects moving at constant speeds, not objects that are accelerating. Acceleration is motion in which either an object's speed or direction (that is, its velocity) changes. Mathematically, acceleration and gravity are equivalent, just like energy and mass. If Angel is riding in the elevator at Wolfram & Hart's high-rise headquarters, and someone cuts the cable, he will go into free fall, feeling as if he were weightless as he floats inside the elevator. Since both he and the elevator are falling at the same rate, he won't be able to feel gravity's pull. So from his limited perspective, Angel might conclude (erroneously) that gravity had inexplicably disappeared. The reverse happens if Angel accelerates in his convertible: He will feel a force pushing him into his seat. If Angel can feel gravity's influence, he can conclude that he is accelerating.

This has important implications for time. Under general relativity, time can flow at different rates for different observers even if they aren't moving relative to one another, provided their respective gravitational fields have different strengths. Einstein envisioned a space-time that is curved, not flat, and hence gravity is not so much a force as it is space-time that has been bent out of shape by the presence of mass or energy. How much mass or energy is present determines the degree of curvature, and the more it curves, the stronger the gravitational pull. Since space and time are one, what affects space also affects time: As space is warped, time is stretched or compressed accordingly. Therefore, time slows down in direct proportion to the strength of a gravitational field, and that field's strength depends on distance.* A 1960 experiment at Harvard University to measure gamma radiation frequencies demonstrated as much. Time passed more slowly for an atomic clock set up in the basement than it did for an identical clock on the roof 74 feet above, simply because the first clock was closer to the Earth's gravitational field.

Einstein's equations opened up an entirely new realm of theoretical possibilities. For instance, it is conceivable that if there is enough mass or energy present, the fabric of space-time can curve so much that it folds back on itself. This creates a temporal fold where the present folds back and touches a point in the past, just like the one Angel experienced. Physicists call these closed timelike curves. Of course, this is primarily a mathematical concept. Scientists have yet to observe a genuine closed timelike curve anywhere in the universe, because the amount of mass or energy required to give rise to one is simply too great for our current meager resources. But general relativity can't rule it out completely.

The ancient demon Illyria also possesses the ability to warp

*This is the inverse square law discussed in chapter 2. If something is twice as far from the source of a given force, like gravity, the strength of that force acting on the object will decrease by four times as much.

the fabric of space-time, jumping into her own personal "hyperdrive" at will. When she does this, from her point of view she is moving at a normal speed, but Team Angel is moving in ultraslow motion, almost frozen in place. They, in turn, perceive Illyria as moving so fast, she is little more than a blue blur, almost beyond their perception. As with Buffy and the time-inhibiting bug, were Illyria wearing a watch while in "hyperdrive," it would mark the passage of time much more slowly than the timepieces worn by Team Angel, because she is moving so much faster. Her unique ability is linked to the energy associated with her demon essence, now housed in a human shell (formerly known as Fred).

In "Time Bomb" (A-5), Illyria's power spirals out of control, and she begins jumping back and forth randomly in time. At one point, she accuses Angel of ripping her out of linear progression: "How do you unweave time this way?" But it is Illyria herself who is responsible. She has begun "leaking" radiation energy. Her human shell can't contain her full demonic power, and is slowly disintegrating. Spike likens it to a crack in her engine block. The radiation makes it easy for Wesley to track her with a Geiger counter, but it also appears to cause time to fold back on itself repeatedly, each fold transporting Illyria to a different point in the established time line.

Just how much energy would be required to bend space-time back on itself? A mere "leak" would be insufficient, but how about a nuclear blast? Eventually Illyria's human shell crumbles completely, and all her pent-up demon energy is released at once. She literally explodes, taking most of Southern California with her. In the process, her chronological time line is blown to bits, and she finds herself at some earlier point in the time line. Since the outcome is always the same, each time she explodes, she is thrown back through time yet again. We might conclude, as Angel does, that the energy of the blast is so powerful, it severely warps the space-time in her proximity. It's strong enough to blow Angel back in time, enabling him to change the tragic outcome and stabilize the chronological time line.

Unfortunately, outside the Buffyverse, even the blast from a thousand nuclear bombs would be insufficient to create a single temporal fold. Our Sun produces enough nuclear reactions (roughly 10^{38} per second) in its core to power a million hydrogen bombs every second, but we would need all the energy in the galaxy put together, and possibly more besides, in order to bend space-time to such a degree as to give rise to temporal folds. Illyria and the Oracle are very powerful beings indeed if they can tap into that vast an energy source so easily.

TIME PLAYS LOOP THE LOOP

In the third and final test designed by the Trio, Buffy finds herself caught in a time loop, another bizarre mathematical anomaly allowed by Einstein's equations. She is doomed to repeat the same sequence of events—trying to satisfy a demanding Magic Box customer intent on purchasing a mummy hand—over and over until she gets it right, much like Bill Murray's beleaguered weatherman character in the film *Groundhog Day*. Time loops are another example of closed timelike curves, simply in a different configuration.

Igor Novikov, a cosmologist at Copenhagen University Observatory, has compared time to a river that flows from the past into the future. Under Einstein's relativity, that river would speed up or slow down as it "flowed" through the universe. But a closed timelike curve essentially separates one small piece of time's river from the main flow of events, like a whirlpool or eddy. Under general relativity, it is mathematically possible to travel in the same direction, always moving forward, and yet still end up right where you began—shades of the Mayan *lamat*. If the presence of mass-energy were great enough (that is, approaching infinity) in a particular area of space-time, it could conceivably become so warped that the elliptical path would come full circle and the two ends would meet, forming a closed loop. Such a thing could account for the time loop that turned Buffy's first (and last) shift at the Magic Box into major overtime. Buffy becomes trapped in a closed timelike curve. Space-time loops back to the same starting point, until Buffy finds a way to meet Jonathan's condition: Satisfy the customer. It's similar to the way a computer program will keep looping in on itself until a certain set of predetermined conditions is met.

Time isn't running backward; it's running in circles. The planets in our solar system actually move in straight lines. Although they appear to move in elliptical orbits, they are merely following the curvature of space-time. It's a bit like driving from Washington DC to Los Angeles. It seems as if we are traveling along a straight path, when in fact we are moving along the curve of the Earth. We just can't see that curvature. Buffy also follows a curved path in space-time. The curvature is simply more severe. If Buffy were to check her watch while looping, she would find that it always runs forward, even though she eventually ends up right back where and when she started.

It's worth noting that Stephen Hawking, for one, has objected to the theoretical possibility of closed timelike curves, not just because of the extraordinary energies required, but because such phenomena violate something called causality. He calls it his "chronology protection conjecture": The laws of physics conspire

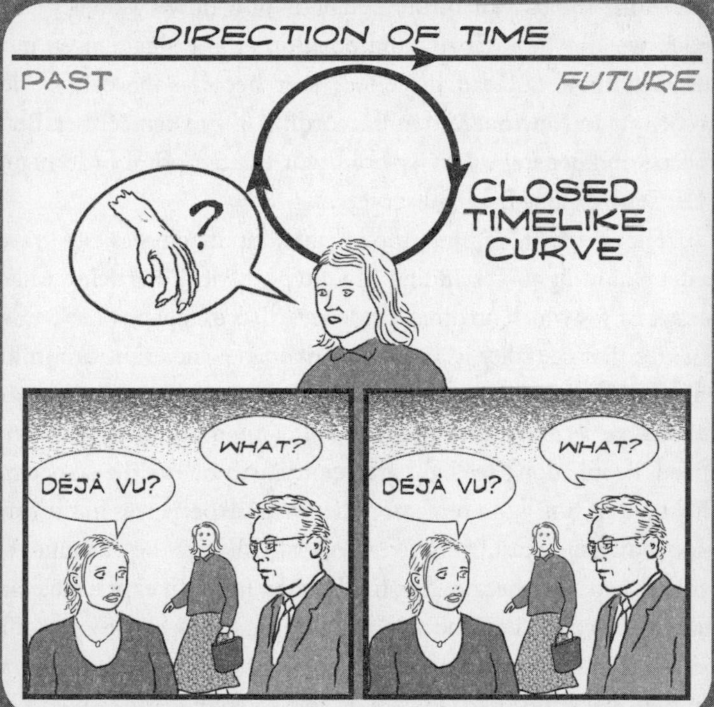
DIRECTION OF TIME
PAST
FUTURE
?
CLOSED TIMELIKE CURVE
DÉJÀ VU?
WHAT?
DÉJÀ VU?
WHAT?

to prohibit time travel to the past—or at least make it highly unlikely—and hence avoid such troublesome paradoxes as accidentally killing one's grandfather before one is born. Every event has a cause, of which it is the effect. Spike hits the cue ball to set it in motion; the cue ball collides with the eight ball; and the energy transfer from that collision causes the eight ball to roll into a pocket of the pool table. Thanks to Newton's equations, we can determine the past or future behavior of a moving object, provided we know its velocity and position at any single given moment in time. But can the effect ever become the cause? Not according to Newton. And not according to Einstein, either. Both special and general relativity hold that the temporal ordering of events is the same for all observers.

The fact that nothing known thus far in physics can travel faster than light—excluding the hypothetical particles called tachyons, for which no direct evidence exists except on *Star Trek*—ensures that causality is always maintained. One event can influence a second event if, and only if, a signal could be sent from the first to the second event, and this can happen no faster than light speed. People don't see light arrive on Earth before the stars emit that light. If Buffy can remember her prior experiences in the time loop—and she must, if she ever hopes to change the outcome—it violates causality, because she has learned from an experience that has yet to occur. The same principle applies to Angel. He is the only person who remembers the day that never happened after the Oracle folds back time and, as such, is able to change the sequence of events to produce a more desirable outcome. Ditto for his temporal adventure with Illyria. Both experiences violate causality. Illyria even says as much, declaring that Angel's presence in her fractured time line is an "impossible paradox."

In a true closed timelike curve, one can only continually relive the same sequence of events, not alter the past, and therefore one cannot change the future. In "Waiting in the Wings" (A-3), a powerful Russian wizard named Count Kurskov, who heads the Blinnikov World Ballet Corps, has the ability to pull his prima ballerina out of time. Angel saw the troupe perform *Giselle* in the 1890s, and when they come to Los Angeles 100 years later, he finds that everything is identical, right down to the dancers and their steps.

Kurskov has trapped his beloved dancer (and the entire company) in a looping mini-universe, where she is doomed to repeat the exact same performance for eternity. As she tells Angel when he finds her, "I don't dance, I echo." Here, causality is maintained (although the ballerina seems to be aware of her unique cyclical existence). The same sequence of events plays out again and again, unaltered—at least until Angel figures out a way to break the spell. The ballerina merely has to dance different steps while the count's power is momentarily weakened. The moment the sequence of events changes, the closed timelike curve is undone and the looping mini-universe rejoins the normal flow of time. The ballerina and her company cease to exist, as they should have done long ago.

TIME STANDS STILL

Closed timelike curves redirect the forward flow of time back into the past, but there are other twisting permutations of spacetime where that flow could conceivably stop altogether. Most men have a hard enough time coming up with the perfect anniversary gift for a woman. Imagine how much more difficult it would be if you know she's going to break up with you afterward. That's the dilemma facing a brilliant physics graduate student named Gene in "Happy Anniversary" (A-2). Gene is working on a method for stopping time, earning him the nickname "Time Boy" from a jealous colleague. "It's not freezing time, although that's how it would look to an outside observer," Gene explains to a lab assistant, demonstrating his grasp of relativity. Instead, he's trying to carve out an infinitesimally small piece of space-time and remove it from reality. Anything contained within that moment would exist in its own little bubble universe, forever unchanged. Like the Oracle's realm, time would have no meaning.

Unfortunately, he can't get the math right. It's a dilemma every physicist has faced: No matter how much you tweak the numbers or your apparatus, the damn thing still won't work. But Gene has something the average physicist lacks: the secret admiration of a technologically advanced, fanatical demon sect that seeks to rid Earth of the pesky human race for good. It turns out that there's a flaw in his system. If the experiment is improperly contained, the tiny bubble universe will spill out and engulf the entire world.

The demons sneak into the lab and correct the equations, intent on exploiting that flaw. When Gene returns, he enters the new data—and the experiment works. But when he overhears his girlfriend confiding her intention to break up with him, he sets up the experiment in his bedroom and triggers the device at the climactic moment. As Gene and his girlfriend are suspended in their private bubble, the demons remove the safeguards and allow that bubble to spread outward into Los Angeles, freezing everything in its path until Angel shuts down the system. The bubble collapses back in on itself, and time resumes for everyone—including the soon-to-be-jilted Gene.

As explained in the episode, Gene uses a laser-driven particle accelerator, small enough to fit on a tabletop in the lab, to create a strong electromagnetic field around a specific area, trapping anything within it. When we hear the term "particle accelerator," most of us either draw a blank or conjure up an image of the enormous atom smashers that hurl elementary particles down giant tunnels, using magnetic fields to focus and steer the particles into head-on collisions with targets. But there are small-scale versions that use ultrafast laser light to focus and speed up beams of electrons to high energies over very short distances: a few centimeters, compared to two miles. They're called wakefield accelerators, and they do indeed fit on a tabletop, although they can't stop time. Very intense, short pulses of laser light are fired into a form of ionized gas called a plasma. Those pulses produce a wave of energy that ripples through the plasma, leaving a wake of charged particles, much like a motorboat racing across a lake churns up a water wake in its path. A second laser pushes even more electrons into the plasma, and these can "surf" the wake of the laser pulse, picking up speed by draining extra energy from the wakefield.

Gene's premise is that if a single drop of mercury is dropped into the field created by his accelerator, and if the particles that make up the laser beams are moving at just the right velocity, the mercury would be accelerated completely out of our space-time. However, even assuming that such a scheme would work, it's doubtful we could get a drop of mercury moving fast enough. To "stop" time, Gene would need to accelerate his particle beams to exactly the speed of light, and position them in such a way that anything within its field also speeds up accordingly. Accelerating something to exactly the speed of light causes time to contract to nothing. But we've already seen that no object with mass—whether it's something as small as a drop of mercury or something as large as Gene—can ever completely reach the speed of light. Even in giant particle accelerators, scientists can only speed up particles to within 99.99% of the speed of light. No matter how much additional power they feed into the machine, because of the corresponding increase in mass, the particles never quite reach light speed. A wakefield accelerator couldn't do any better.

Extreme gravity, however, might stop time indefinitely, at least from the perspective of an outside observer. The most likely source of extreme gravity is a black hole, which forms when a massive star has used up all its fuel. The reason the Sun and other stars emit light is because trillions of nuclear reactions are taking place at their cores. With core temperatures of millions of degrees, hydrogen atoms can convert into helium atoms, emitting radiation in the process. At some point, however, all the atoms are used up and no more nuclear fusion can take place. Without that outward counterforce to the pull of gravity, a star collapses inward, eventually reaching a point where the attractive gravitational force is so strong, not even light can escape.

No one has ever observed the center of a black hole; until quite recently, such objects only existed in theory. But scientists surmise that a black hole has at its center an infinite density and an infinite gravitational field, as well as infinite entropy, which means that no further change can take place. No physical processes can occur, and so time can be said to have stopped. This point is known as a "singularity." Black holes—and their extreme gravity—do seem to be a factor in Gene's research. At one point, he compares the effect of his temporal experiment to a "tiny event horizon." The event horizon of a black hole is not so much a physical surface as the theoretical point of no return for any object that gets caught in the black hole's powerful gravitational field.

For the sake of argument, let's say that Gene's small tabletop apparatus is powerful enough to produce large enough energies at sufficient densities to create a small black hole in the laboratory, even though real-world physicists have yet to do so in a full-scale particle accelerator. What would happen to Gene as he moved closer and closer to the black hole's event horizon? To an outside observer looking through the lab's protective glass, Gene would move ever more slowly the closer he got to the event horizon.

Eventually the gravitational field would become so strong, and time would move so slowly, that Gene would appear to be frozen in time, just at the edge of the black hole. He would be on the verge of falling into it for eternity.

Alas, from Gene's perspective, the view isn't quite so rosy. Once he crossed the event horizon, he would be sucked inexorably toward the center of the black hole. There would be a vast difference between the gravitational forces on his head and on his feet: the force on his feet would always be greater than the force on his head. So Gene would stretch out like spaghetti. Ultimately he would be torn apart long before he reached the black hole's singularity. Time wouldn't just stop for poor Gene: it would end completely.

IMAGINARY TIME

Yet there could still be hope for our lovelorn physicist. For decades, physicists assumed that the mass of a black hole, and the radius of its event horizon, remained constant. General relativity (which applies to the macroscopic world) demands it. However, this doesn't take into account quantum mechanics, which governs the world of tiny subatomic particles: atoms and the bits and pieces that comprise them. The equations Gene scribbles on his blackboard come from quantum mechanics, so we can safely assume that it also plays a significant role in his time-freezing research. It turns out that black holes aren't as black as scientists originally thought. They can and do emit tiny particles of radiation, and this emission causes a black hole to lose mass, until it eventually winks out of existence, albeit on a cosmic time scale.

This is a direct consequence of quantum fluctuations, a

peculiarity in which the established conservation laws of physics can be violated—but only on time scales so small, they are nearly instantaneous, and thus beyond our ability to observe them directly. We'll learn more about the strange world of quantum mechanics in the next few chapters. For now, it's enough to know that on occasion, empty space (known as the quantum vacuum) can spontaneously produce pairs of "virtual" particles, each with the same mass, but opposite charges, so they annihilate each other almost immediately. So empty space isn't empty at all. It is boiling with virtual particles that continuously wink in and out of existence. If a particle pair pops out near a black hole, instead of annihilating each other, one may fall in while the other escapes and is emitted as radiation. Energy appears to come from nothing. The mass of the black hole must decrease ever so slightly in response to account for it, so that energy is still conserved.

Once a black hole evaporates, what happens to all those objects that fell into it? According to Hawking, one possibility is that these objects go into a little baby universe all their own: a small, self-contained region that branches off from our own universe, much like Gene described as the hoped-for outcome of his experiment. Anything sucked into a black hole would not survive, including tiny elementary particles. But those particles could, hypothetically, cross over into a baby universe, reemerging much, *much* later as particles emitted by another black hole.

This neat little trick is possible thanks to a concept Hawking has dubbed "imaginary time." It's related to the concept of imaginary numbers, which result when one tries to learn the square root of a negative number. Gene no doubt learned in high school math class that you can't determine the square root of a negative number, because when you multiply two negative numbers, you always end up with a positive answer: -2×-2 will always equal 4. But if you start out with an imaginary number, the square will be negative by definition: The square root of –4 is –2*i*. Imaginary numbers enable physicists like Gene to plot time as though it were just another dimension of space. And that's essentially what imaginary time is: a second temporal dimension. After all, if space can have three (or more) dimensions, why can't time have more than one?

It's best to think of imaginary time as a direction of time that runs at right angles to real time. If we envision that real time runs as a horizontal line (one dimension), with the past on the left and the future on the right, then imaginary time would run vertically, up and down. Time now has two dimensions. Real time is linear, moving inexorably forward in only one direction, thanks to entropy, but imaginary time has length and breadth, and therefore a kind of surface area. It can stretch out infinitely in all directions, just like two-dimensional space.

This is critical, because it gets rid of that nasty, bone-crunching singularity at the center of a black hole, where everything collapses down to an infinite point in space-time. Instead of a single, linear, chronological time line through three-dimensional space, we get what amounts to a four-dimensional sphere. The closest analogy from our limited human perspective would be the earth. Say we begin our temporal journey at the North Pole and then progress all the way to the equator. Real time will run in one direction as a straight line from the starting point to an equatorial endpoint, but imaginary time will spread out across the sphere in all directions, much like casting a two-dimensional net over the earth. There is nothing particularly special about the North Pole, other than that we chose it as our starting point. We could cast the same sort of "temporal net" from any random point on the globe. There are no sharp edges, no boundaries, to a spherical universe in imaginary time, and hence no quantum singularities.

Let's be clear: These two kinds of time are separate entities. In real (linear, or one-dimensional) time, Gene would still be torn apart by the intense gravity of the original black hole as he fell into it, and the particles that make up his body would come to an end at the singularity. A one-dimensional line will always collapse into a point. But in imaginary (two-dimensional) time, the particles would continue, because the singularity doesn't exist; a two-dimensional structure collapses into a line. It's a bit of a mind-boggle, but let's pretend that the vertical (imaginary) axis splits the earth into two separate hemispheres, east and west. Gene's black hole exists in the western hemisphere, but there is another black hole in the eastern hemisphere—that is, on the other side of the original black hole's singularity, which serves as a bridge between the two.* Gene's particles would pass through the singularity—moving from the western to the eastern hemisphere—and reemerge as particles emitted as radiation by the second black hole. Real time would end at the singularity, but imaginary time would continue on the other side. Of course, this journey would occur on cosmological time scales, and the particles that reemerge would not be the same particles, nor would they be in the same form as the original object. So this still might not be a desirable outcome for Gene.

Imaginary time might seem like an unnecessary complication. Hawking proposed it, in part, because our understanding of time is incomplete. Contradictions arise because relativity and quantum mechanics are vastly different theories: one applies to the physics of the very large, while the other describes the physics of the very small. Black holes—the apparent key to Gene's time-stopping experiment—are bizarre anomalies that encompass both, since their singularities are infinitely dense and infinitely small at

*This is known as an "Einstein-Rosen bridge," and will be discussed in more detail in chapter 9.

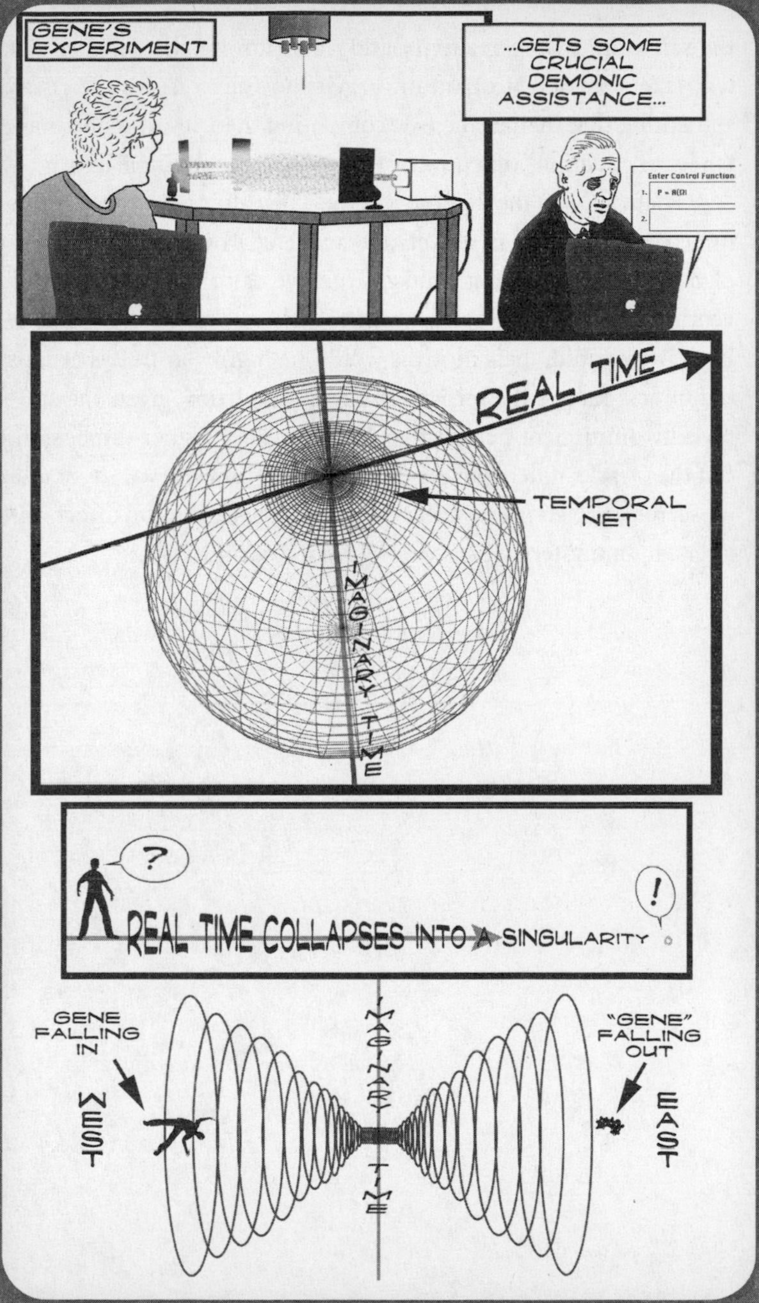
GENE'S EXPERIMENT
...GETS SOME CRUCIAL DEMONIC ASSISTANCE...
Enter Control Function
1. P = A(Ω)
2.
REAL TIME
TEMPORAL NET
IMAGINARY TIME
?
REAL TIME COLLAPSES INTO A SINGULARITY
!
GENE FALLING IN
WEST
IMAGINARY TIME
"GENE" FALLING OUT
EAST

the same time. Both relativity and quantum mechanics apply. It will take a theory of quantum gravity to fully resolve the issue, something that merges the two competing theories into one master set of equations that govern the universe, including time.

That could be the demons' secret: They alter Gene's equations in accordance with an as-yet-undiscovered (by humans) theory of quantum gravity. But unlike Gene, we can't rely on the intervention of a fanatical demon sect to help us out of the muddle. Even if we could, their motives would be highly suspect. Perhaps the Oracle knows the answer to the conundrum, given the supposedly omniscient perspective of its timeless higher dimension. But the Oracle, much like the universe itself, doesn't willingly spill its secrets, preferring to offer cryptic clues in lieu of direct answers. It's a mystery we have to figure out for ourselves.

6

IMPROBABLE CAUSE

When Quantum Weirdness Reigns Supreme

"Apparently I did not observe myself properly when I was in a superposition of states just now."
—Alice, ALICE IN QUANTUMLAND

High school can seem like hell to teenagers, especially those who are a little bit different and don't quite fit in with the "popular crowd." It's even worse in the Buffyverse, where high school can quite literally be hell. The students at Sunnydale High have the misfortune to attend classes atop a Hellmouth, whose mystical energy has the potential to warp reality, often with spectacular consequences. In "Out of Mind, Out of Sight" (B-1), a shy Sunnydale student named Marcy mysteriously turns invisible, then uses her newfound power to seek revenge on everyone who has ignored or mistreated her.

A baffled Giles finally stumbles on an explanation for Marcy's bizarre condition, drawn not from his usual books on mysticism, but from physics—specifically, quantum mechanics. It's a clever twist on a peculiar attribute of the quantum world: There is a sense in which reality can be shaped or created by observation—or, in Marcy's case, by the lack thereof. Marcy flew under everyone's radar, unnoticed for months by students and teachers alike—she wasn't even deemed worth bullying—and this finally manifested itself physically. People perceived Marcy as invisible, and she became so.

The notion that perception dictates reality is a theme that recurs frequently in the Buffyverse. In "Nightmares" (B-1), people's bad dreams begin to manifest themselves until the nightmare world threatens to overwhelm reality completely. "We are defined by the things we fear," the Master tells one of his minions. That's certainly true for the college kids who attend a haunted house frat party in "Fear, Itself" (B-4). A Celtic fear demon named Gachnar draws energy from the frightened partygoers, enabling the demon to make everyone's worst fears tangible. And sometimes you really are what you wear. Ethan Rayne casts a spell to turn trick-or-treaters into their costumes in "Halloween" (B-2).

Odd as these occurrences may be, quantum physics can be even stranger. It seeks to describe nature at the level of individual atoms and the particles that comprise them. But when physicists began delving into the strange new realm of subatomic particles in the early twentieth century, they discovered a world where the old, deterministic laws of classical physics no longer apply. Instead, uncertainty reigns supreme. It is a world governed not by absolutes, but by probabilities, where events that would seem impossible on the macroscale occur on a regular basis. In the Buffyverse, the strange properties of quantum physics provide a treasure trove of innovative concepts that the writers used to create all kinds of fictional quantum quandaries.

THE HELL GOD HAS TWO FACES

In Season 5 of *Buffy*, an exiled hell god named Glory arrives in Sunnydale with her scabby minions and exhibits some peculiar characteristics. To ensure that Glory stays in exile, she has been trapped inside the body of a human male: a kindly young doctor named Ben at Sunnydale's hospital. But a mere mortal like Ben is unable to entirely contain a being as powerful as Glory. At random moments, she reemerges, morphing into a comely young woman and wreaking all manner of havoc. When her power weakens, Ben is able to resume control of the body.

Although the Scoobies repeatedly encounter Ben and Glory, they never see both personas present at the same time, and hence don't suspect any connection between the two. And thanks to a convenient magic spell, should any humans witness the transformation, they will promptly forget they saw anything. So which identity is the "real" one, Ben or Glory?

The answer is that both are real. While one identity is always subsumed into the other, the dominant persona switches back and forth. The same is true for subatomic particles, which also have two identities—particle and wave—that take turns manifesting in nature. Light, for example, is made up of collections of tiny particles, called photons. It travels in a straight line, can't turn corners, and bounces off a mirror. Yet light can also spread outward, like a wave, and possesses the wavelike properties of frequency and wavelength. The technical term is complementarity: Waves and particles are complementary aspects of the same phenomenon. Photons, electrons, protons, and even neutrons all exhibit wave/particle duality. Which identity is present at any given moment turns out to depend on what we're trying to measure.

We learn in "Selfless" (B-7) that Willow is taking a physics class at UC-Sunnydale. It's likely that Willow would be required to replicate one of the most famous experiments in physics as part of her lab work. Let's say that Willow sets up a light source in front of a barrier that has two small slits cut into it. On the other side is a light-sensitive screen to record the pattern of incoming light. She turns on the light source and sends one photon at a time toward the two slits in the barrier. A single photon should only be able to go through one slit or the other. Instead, she finds that the light forms a pattern of alternating bands of dark and light on the other side. This is known as an interference pattern, an indication of wave behavior.

Recall from chapter 1 that whenever the crests and troughs of sound waves line up, their amplitudes add together, producing more decibels (volume). The same thing happens with light waves. Wherever their respective crests are perfectly aligned, the combined amplitudes form a bright spot on the screen in Willow's experiment. Wherever the crests of one wave line up with the troughs of the other, the two waves cancel each other out, leaving a dark spot on the screen. Willow can conclude that single particles are mysteriously behaving as if they were waves. Each photon has somehow managed to travel through *both* slits and interfere with itself on the other side. Something even stranger happens when Willow places particle detectors by the slits in an attempt to verify that the photons go through both openings. The interference pat-

tern disappears. Now the photon is behaving like a particle, passing through one slit, or the other, but not both. Its wave identity is subsumed. Somehow the photon always "knows" what to do, behaving as a particle or a wave as the occasion dictates.

There's a simple explanation for Willow's seemingly contradictory results, albeit one that defies common sense. The first version of the experiment is a wave measurement; the second version is a particle measurement. The photons aren't making a conscious decision to behave one way or the other. Rather, Willow's making a specific measurement forces either the particle or wave feature to be dominant. When Willow adds the particle detectors, she inadvertently changes the nature of the experiment, and hence the outcome. If she allows the photons to travel from light source to screen unobserved, they behave like waves, producing the interference pattern. But if she observes them en route, she knows which path the photons took, and this forces them to behave like particles.

Thus, as Giles maintained (and the unfortunate Marcy could confirm), quantum mechanics appears to dictate that even the most objective of experimental outcomes is dependent upon human observation. Set up an experiment with a wave detector in place, and light will exhibit wavelike behavior. Repeat the same experiment with a particle detector in place, and light will show up as a particle. Willow can construct her experiment to produce an interference pattern, or to determine which way the photon went. She can't do both at the same time, any more than she can see both Ben and Glory in the same place at the same time.

It all boils down to a fundamental principle called uncertainty, first postulated by the German physicist Werner Heisenberg. The uncertainty principle serves much the same purpose as the magic spell that protects Ben/Glory's dual identity. It holds that we can know the precise momentum (or the precise velocity) associated with a particle, or we can know its exact location, but we can't determine both at the same time. Those properties are mutually exclusive, because the very act of making a measurement changes the state of the atom, much like the Scoobies' behavior changes when Andrew videotapes them in "Storyteller" (B-7): They self-consciously "play" to the camera when they realize they are being filmed.

Subatomic particles also alter their behavior once they "realize" they are being watched. It's not magic, or even a Zen-like paradox: The change is the result of an actual physical force. If we want to bounce light off an atom in order to determine its location, we must shine light with a wavelength small enough to be disturbed by the atom. Light only comes in small packets (or quanta) of energy, called photons, and a photon's energy is inversely proportional to its wavelength. The shorter the wavelength of light, the higher the energy of its photons. The smaller the object we wish to observe, the higher the energy of the light we must use in order to achieve sufficient resolution. An atom is very, very small. If we wish to locate an atom's precise position, we must bombard it with a photon of such high energy that significant amounts of energy will be transferred to the atom itself, changing its speed or direction.

If the mere act of observation can determine the outcome of an experiment in our world, and if a continuous lack of observation turns Marcy invisible in the Buffyverse, what happens if we never look away? The series takes no stance either way. However, according to quantum theory, experimental measurements are made in single, fixed brief moments in time, but if an experiment is observed continuously, time effectively stands still.

This is a quantized version of something called Zeno's paradox, which dates back to ancient Greece. Zeno used it to logically construct an argument that proved the nonsensical assertion that motion is impossible. Transposed into the world of the Buffyverse, it goes something like this: If Buffy fires a wooden stake from her crossbow, the stake will travel in a straight line indefinitely until it is stopped by an opposing force—say, the heart of an approaching vampire. But what if you could divide the distance the stake must travel to its target into an infinite number of increasingly smaller increments, halving the distance every step of the way? The stake would get closer and closer to its target, but would never be able to reach the vampire's heart. All motion, and therefore time, would seem to stop.

Of course, this simply doesn't happen. Eventually, Buffy's stake will find its mark and the vampire will explode into dust. That's because Zeno's abstract argument rests on the assumption that the progression will go on for infinity. But in physical reality, there are no infinities; there is always some kind of limit. An endless series can have a finite sum. The distance Buffy's stake must travel would eventually reach a point where it could no longer be divided, because those increments would be smaller than the stake itself. At that point, the stake would reach the vampire's heart.

Here's the twist: On the subatomic level, something very similar to Zeno's paradox really happens. In 1977, researchers found that a radioactive atom would never decay if it were observed continuously, which they dubbed the quantum Zeno effect. Continuous measurement ("observation") apparently has the same effect on atomic energy levels as dividing the distance traveled by Buffy's stake into an infinite number of ever-smaller increments. This has been experimentally confirmed. In 1989, a team of scientists at the National Institute of Standards and Technology (NIST) in Colorado trapped 5,000 charged beryllium atoms in a magnetic field and then tried to "boil" them by zapping them with a radio frequency field to raise their temperature. They expected the atoms to absorb the extra influx of energy and jump to higher ("hotter") energies. But this only occurred if no measurements were made in the interim. The more often they tried to measure

THWAK!
$\exists p\ \forall i\ L(p, i) \rightarrow R$
MAN BITES WEREWOLF
VAMPIRE TIMES
GRRRR
$\exists p\ \forall i\ L(p, i) \rightarrow R$

the energy state of the atoms, the fewer atoms would reach the higher energy level. At the rate of one measurement every four-thousandths of a second—essentially continuous observation—no atoms at all jumped to the higher energy state. They simply refused to heat up. The effect occurs even when an automated measuring device is used.

The culprit is uncertainty. The very act of measurement interferes with the atoms' ability to absorb extra energy. By definition, the energy state of an atom moving between two energy levels (the lower ground state and a higher one) is a bit fuzzy. Uncertainty is a very abstract property, so it helps to think of it in more concrete terms, such as a stake fired from Buffy's crossbow. This particular stake is endowed with the mysterious ability to grow longer over short intervals of time. That's what happens to the uncertainty associated with two atomic energy states: It grows larger. When uncertainty becomes large enough to bridge the two energy states—akin to lengthening Buffy's stake to the point where it can reach the vampire's heart—the atom shifts to the higher energy state (and Buffy's stake dusts the bloodsucker). The "uncertain stake" then collapses back down to its original length and the whole process starts over again.

The catch is that every time we make a measurement of an atom's energy, it reduces uncertainty. Imagine if every time Buffy measured her stake to determine its length, it became just a little bit shorter, until it simply wasn't long enough to bridge the remaining gap between itself and the vampire's heart. The stake would never meet its mark. That's what happens to the energy states of atoms in the quantum Zeno effect. Uncertainty gets smaller with every measurement, because each measurement yields new information about the atoms, reducing the "fuzziness" of their energy states. Make those measurements often enough, and uncertainty never becomes sufficiently large to enable the atom to jump to a higher energy level. So a "watched" quantum pot never boils.

BEATING THE ODDS

Uncertainty is often misinterpreted outside scientific circles to imply that nothing certain can be known about the quantum world. Such is not the case. The subatomic realm follows its own set of physical laws that, in their own way, are as deterministic as the laws of classical physics. The problem, as we have seen, is that we can't measure the properties of quantum objects precisely, since the very act of measuring them interferes with the objects, altering their properties in the process. So we can only determine the probabilities of particles in any given state at any given time.

Any casino in the Buffyverse is most likely run by demons or other evil beings, and therefore doesn't play fair. Nevertheless, Fred, a former physics student with a head for numbers, manages to amass a respectable stack of chips at the blackjack tables in a Las Vegas casino by counting cards in "The House Always Wins" (A-4). She is able to do this because the statistical probabilities of blackjack are well-known, and can be easily calculated by someone with her stellar mathematical abilities.

For instance, if Fred is dealt a face card on the initial pass, she knows there is a 1 in 13 chance of drawing an ace and making 21 on her second card. Unless she cheats and sneaks a peek at the cards themselves, she can't say for sure that she will draw an ace. Each remaining card in the deck carries its own set of probabilities. The more cards there are, the tougher her calculations become—which is why most casinos use multiple decks for blackjack. But she can determine when the odds are in her favor and when they are not, and place her bets accordingly to come out ahead in the end.

Quantum physicists face the same kind of challenge when calculating probabilities for subatomic particles. But the problem is compounded in the quantum world by the huge number of particles involved in even the most basic physical process. It's akin to Fred having to calculate her blackjack odds using billions of inter-shuffled decks of cards, which would tax even her mental prowess. Consider the reflection of light from a mirror, which is comprised of billions upon billions of atoms. If a steady stream of photons hits the surface of a mirror, for every 100 photons that hit the glass, 95 are reflected, and 5 are lost, passing through or being absorbed by the glass. But physicists can't predict which specific photons will be reflected or lost, any more than Fred can say for certain that the next card in a deck of billions will definitely be an ace. They can only say with certainty that out of every 100 photons, 95 will be reflected and 5 will be lost.

Nonetheless, there are some subtle but critical differences between the statistical predictions of classical and quantum physics. Classically speaking, if Fred flips a coin, she can say that there is a 50/50 chance it will land heads or tails, simply because she doesn't have enough information about the many factors that could influence its landing position—the rate of spin, height, or a slight gust of wind, for example. It is, in principle, possible for Fred to acquire more information to refine her prediction so that it is more accurate. It's not even an option in quantum physics. No further information is available beyond the stated probability. We have uncertainty to thank for that. The act of acquiring more information about subatomic particles irrevocably alters them, so the information we gain is no longer valid, and we are right back where we started.

The other critical difference has to do with an object's state when we aren't looking at it. Under the classical physics model, prior to Fred's "measurement" to determine how the coin fell, we know that one side is definitely heads and the other is definitely tails. So there is a 50 percent probability that the coin will land heads, a 50 percent probability that it will land tails, and zero probability that it will be both heads and tails at the same time.

The same cannot be said of the quantum world. It constitutes a reality in which things can hover in a fuzzy, nebulous cloud of probability that encompasses all potential states: heads and tails, particle and wave, Ben and Glory. Things become definite only when an observation forces them to settle on a specific outcome. Technically, the coin exists in an indeterminate state of both heads and tails until it lands and Fred can see which way it fell. Similarly, if Fred's blackjack cards were subatomic particles and she wasn't counting them (counting would constitute a measurement or observation), all possible cards would be present at all possible times. A card would only take on a definite "identity" when it was flipped over to reveal its face.

This bizarre superposition of states is described by a mathematical equation called the wave function. The brainchild of Austrian physicist Erwin Schrödinger, the wave function is simply a formula that tells scientists the probability of finding a particle in any given location in space and time, represented by the Greek letter *psi*. For example, an electron orbiting an atom doesn't have a specific location before it is measured. It exists in a cloud of probability containing all possible outcomes of that pending measurement. The electron is most likely to be found in the area where the cloud is densest, and the least likely to be found where there is hardly any cloud at all. It is only when an object is measured or observed that the wave function "collapses" into a definite state. Ben and Glory can be said to exist in a superimposed state, taking on one or the other identity only when someone observes him/her. Marcy also encompasses all possible states. She is not being "measured," so her wave function hasn't collapsed. She inhabits all possible states and is therefore (at least in the Buffyverse) invisible to human perception.

Fred beats the odds at blackjack in hell's casino, but her fellow gamblers aren't so lucky. All the other games are magically rigged to ensure that the house always wins. Yet Angel hits the jackpot at a slot machine. This highly unlikely outcome is the result of Cordelia's intervention during her brief stint as a Higher Being—an exceedingly rare occurrence in its own right. But it could just as easily be attributed to yet another quirk of quantum mechanics: All outcomes that can happen eventually will happen, given enough time, even if the probability of a certain outcome occurring is infinitely small. In a subatomic casino, the house *almost* always wins. There is the tiniest probability that the house might lose.

Physicists might be able to perform feats indistinguishable from magic if they could only figure out how to control those probabilities. For instance, subatomic particles can "tunnel" through seemingly impenetrable energy barriers. Imagine that an electron is a water wave trying to surmount a tall barrier. Even if the wave is shorter than the barrier, there is a small probability that it will seep through to the other side. This neat little trick has been experimentally verified many times. In the 1950s, physicists devised a system in which the flow of electrons would hit an energy barrier and stop. Most of them lacked sufficient energy to surmount that obstacle. But some electrons didn't follow the established rules of behavior. They simply tunneled right through the energy barrier.

Lorne is being held prisoner in one of the casino's dressing rooms. With the door closed, we can't say precisely where Lorne would be at any point. He would exist in a fuzzy cloud of potential locations, and he wouldn't take on a definite location until the guards checked to see exactly where he was. Still, we could safely assume (as his guards do) that he would always be found somewhere in the room. But in a quantum world, Lorne merely has the highest likelihood of being found somewhere in his dressing room. There is the tiniest probability of finding Lorne on the street just outside the casino. If he waits long enough, eventually he can escape his captors with no effort at all, thereby violating the laws of classical physics, not to mention common sense. Of course, he would have to wait longer than the entire lifetime of our universe before this occurred, but the probability is definitely there.

The same thing happens when an electron is confined in a box, like Lorne confined in his dressing room. Its wave function can seep through the box into the outside world, so there is a small probability that the electron will tunnel through the wall of the box and emerge on the other side. But the electron doesn't have to wait the lifetime of the universe for this to happen, thanks to the vast number of subatomic particles involved at quantum scales. The probabilities that one particle would slip through are much higher. In fact, the phenomenon occurs all the time. So if Lorne were a subatomic particle instead of a big green empath demon, he could escape his captors much more easily.

SPOOKY ACTIONS

It's not just probability that's weird in the quantum world. Sometimes you get two-for-one specials, a phenomenon that also pops up in the Buffyverse. In "The Replacement" (B-5), Xander is accidentally split in two when he is hit by a flash of mystical energy from a demon's enchanted staff. He separates into the strong and weak parts of his personality. Each Xander is physically identical to the other, right down to the individual freckle; but where one Xander is goofy, insecure, and inept, the other is confident, competent, even suave. Because the sum total of Xander's essence is split between them, neither can exist without the other, although they appear to be completely independent entities. Once the Scoobies realize what has happened to Xander, they can encounter one twin, determine that he embodies the weaker personality traits, and instantly conclude that the other twin must possess the stronger traits.

Similarly, when subatomic particles collide, they can become invisibly connected, though they may be physically separated. Even at a distance, they are inextricably interlinked and act like a single object. So knowledge about one partner can instantly reveal knowledge about its twin. If you measure the state of one, you will know the state of the other without having to make a second measurement, because the first measurement determines what the properties of the other particle must be as well. This is called entanglement or, as Einstein preferred to call it, "spooky action at a distance."

It's rather apt that Xander's divided self is the result of a spell, since Cornell University physicist N. David Mermin has described entanglement as "the closest thing we have to magic." Think about it: Disturbances in one part of the universe can instantly affect distant other parts of the universe, mysteriously bypassing the ubiquitous speed-of-light barrier. It defies common sense, and yet it's true. In 1997, Swiss researchers entangled two photons and then sent them flying in opposite directions down fiber optic lines to detectors nearly seven miles apart. When they measured the properties of one photon, the observation produced an instantaneous effect on the other.

Subatomic particles collide all the time, so why is entanglement a relatively rare phenomenon? There are lots of different ways particles can become entangled, but in every case, both particles must arise from a single "mother" process. It's a bit like how identical twins emerge from a single fertilized egg, sharing the genetic material. For instance, passing a single photon through a special kind of crystal can split that photon into two new "daughter" particles. We'll call them "green" and "red." Those particles will be entangled. Energy must be conserved, so both daughter particles have a lower frequency and energy than the original mother particle, but the total energy between them is equal to the mother's energy. We have no way of knowing which is the green one and which is the red. We just know that each daughter photon has a 50/50 chance of being one or the other color. But should we chance to see one of the particles and note that it is red, we can instantly conclude that the other must be green.

In much the same way, the blast from the demon's staff constitutes a mother process that splits Xander into two separate but related entities that share the properties of the original Xander. The original collection of personality traits is divided between the two new Xanders, weak and strong, but the sum total is the same. Since we know that each twin has a 50/50 chance of embodying the weak or strong personality traits, once we determine the traits of one twin, we can deduce those of the second. When the spell is broken, the two Xanders recombine with no ill effects.

Xander's inadvertent split personality also sheds some light on the inner workings of teleportation. Anya is a former vengeance demon. Such creatures are able to teleport, instantaneously moving from one location to another, even across dimensions. But Anya's teleportation powers are severely curtailed when she incurs the wrath of her demon boss, D'Hoffryn, who restricts her to teleporting only on official vengeance business.

Teleportation has become a staple of popular science fiction, so it's not surprising to find a version of it in the Buffyverse. It typically involves dematerializing an object at one point and sending the details of its precise atomic configuration to another location. This information can then be used to construct an exact replica from atoms, arranged in the same pattern as the original. Teleportation eliminates the need to travel through space and time. Anya, or any other object, is transported to any location instantly without actually crossing a physical distance.

All vengeance demons in good favor with D'Hoffryn can teleport at will, and Willow and Tara cast a spell that magically "teleports" Glory into the upper atmosphere above Sunnydale in "Blood Ties" (B-5). It's not quite that easy in the real world. For decades, physicists assumed that teleportation wasn't possible at all because it violated the uncertainty principle. It's something of a catch-22. In order to teleport an object, we must scan it to obtain precise information about its atomic structure. However, according to uncertainty, the more accurately an object is scanned, the more it is disturbed by the process of being scanned. We can't measure a particle without altering it in some way. So how could we extract all the information we would need from an object in order to create an exact copy in another location via teleportation?

In 1993, an IBM physicist named Charles Bennett and his colleagues proposed using entanglement as a mechanism for teleporting information about a particle that could then be used to re-create it in a new location. Their method uses three particles: the particle to be teleported (A) and an entangled pair of particles (B and C). Following these three particles through their myriad interactions is a bit like watching a street game of three-card monte: Blink, and you miss a critical moment. First, B and C are entangled and sent to separate locations. B then interacts with A, and A's information is transferred to B. Since B is still entangled with C, any information transferred to B is also transferred automatically to C, without any need to transmit that information across physical space-time. C essentially turns into A, in the new location.

As always, there is a catch. Physicists can "outwit" uncertainty, but they can do so only once. Their ability to teleport objects is limited by the laws of physics, not by D'Hoffryn's edict, but the end result is similar. Quantum physics demands that the original object be destroyed in the teleportation process. That's because when B scans A, the latter's properties are permanently altered by the interaction. A no longer exists in the exact same state as it did. C is now the only particle in that original state. So only one replica of the teleported object is possible. When Anya teleports herself to Brazil for a spot of vengeance, information about her atomic structure must be passed in some way to the desired location, and a replica of herself is created, while the original is destroyed. So every time Anya teleports, she must destroy her physical self and reassemble an entirely new self at the new location—a gutsy undertaking, even for an immortal demon.

The first experimental demonstrations of quantum teleportation occurred in 1997, when a research group in Austria "teleported" a single photon (technically, information about that photon) across a tabletop, re-creating an exact copy on the other side. As predicted, the original photon was destroyed in the process. By 2003, the technique had been sufficiently developed that scientists at the University of Geneva in Switzerland managed to teleport photons a distance of 1.2 miles through fiber optic cable. The technique has already found a real-world application in

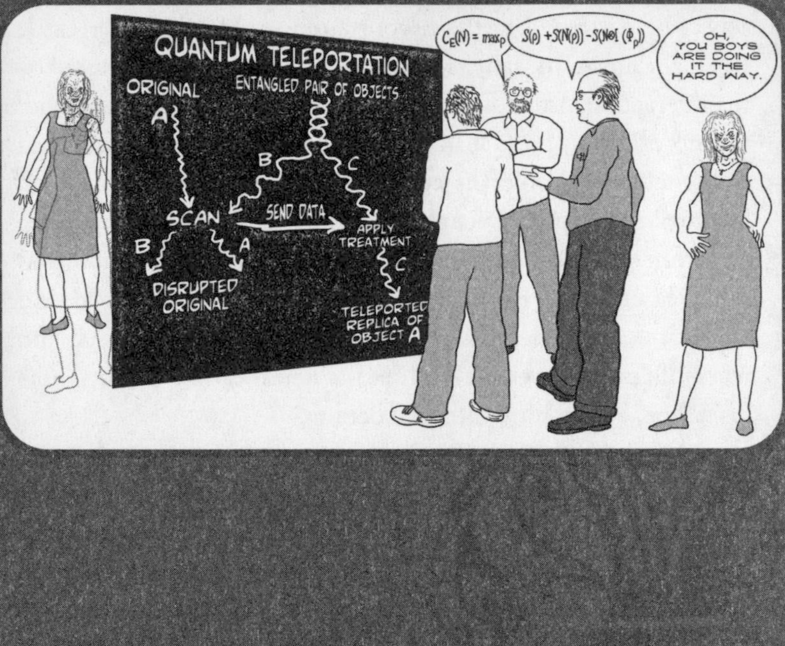
QUANTUM TELEPORTATION
ORIGINAL
A
ENTANGLED PAIR OF OBJECTS
B
C
SCAN
SEND DATA
APPLY TREATMENT
B
A
DISRUPTED ORIGINAL
C
TELEPORTED REPLICA OF OBJECT A
$C_E(N) = \max_\rho$
$S(\rho) + S(N(\rho)) - S(N \otimes I (\Phi_\rho))$
OH, YOU BOYS ARE DOING IT THE HARD WAY.

so-called quantum cryptography: The use of entangled photons as a secure method of communication between two distant parties. Not only is there no solid copy of the "messages" being transmitted, and thus no chance of them being intercepted, if someone tries to eavesdrop on the data stream, it constitutes an observation or measurement. This alters the photons' quantum states and alerts the two parties that their communication channel has been compromised.

Early in 2006, scientists in Tokyo and England made headlines when they managed to teleport not just individual photons, but whole laser beams. And they generated not one, but two near-identical copies of the original beam in a process that has been dubbed "telecloning." Laser beams are "entangled" to connect three points—one "sender" and two "receivers"—and the information about the original is transmitted as in standard quantum teleportation, creating two "copies" that are not quite identical to the original, thanks to uncertainty. Perhaps the spell that divided Xander is the Buffyverse's own form of telecloning. The biggest difference is that real-world telecloning would produce two Xanders with only minor differences, rather than splitting his good and bad personality traits into equal halves. Neither would a telecloning machine be able to recombine the two Xanders to resolve the anomaly.

The possibility of mastering the process sufficiently to teleport a human being (or a vengeance demon) remains in the very distant future. That's because only the entangled particles instantly "know" their relationship. Scientists still have to discern the precise state of two particles by measuring them at either end, and then communicate the information to each other via a phone call or e-mail. This clunky, time-consuming process isn't as much of a problem when teleporting a single particle. But the average human body contains roughly 10^{28} atoms, or more than a trillion

trillion. That's a lot of information to transmit to the new location. Using the fastest fiber-optic system available, it would take the entire age of the universe to transfer all that data.

COSMIC CONNECTIONS

The principles of entanglement are further delineated in Season 4 of *Angel*, when a higher power named Jasmine arranges to be born into a human vessel, intent on bringing about world peace through a form of mass hypnosis. Jasmine has the unique ability to hold in thrall every human being who lays eyes on her. Eventually the entire population of Los Angeles is connected to her. The resulting entanglement is so complete that Jasmine can use the bodies of her followers to see what is happening far from her headquarters. When those followers suffer wounds, the same wounds manifest briefly on her body—her own version of spooky action at a distance.

Before any object can become entangled, or be teleported, it must possess a property called coherence. For example, the photons in a laser beam all vibrate together; they are said to be coherent, although the photons are not entangled. Something similar happens to atoms in a form of matter called a Bose-Einstein condensate (BEC), which was mentioned in chapter 3. In the 1920s Einstein and the Indian physicist Satyendra Bose predicted the possibility that the wavelike nature of atoms would allow them to spread out and overlap if they were packed closely enough together. At normal temperatures atoms behave like billiard balls, bouncing off one another and any containing walls. Lowering the temperature reduces their speed. If the temperature gets low enough (billionths of a degree above absolute zero) and the atoms are densely packed enough, the individual matter waves will be able to "sense" one another and coordinate themselves as if they were one big "superatom." In much the same way, Jasmine's followers can be viewed as millions of subatomic particles maintained in such a strongly coherent state that they think and act as a single entity. But unlike the individual atoms in a BEC, Jasmine and her followers truly seem to be entangled.

Theoretically, many particles should be able to become entangled, provided they all arise from a single, coherent mother process. Indeed, in 2001, researchers in Denmark succeeded in entangling a pair of gas clouds that contained about a trillion atoms each, separated by a few millimeters. The catch is that any entanglement only lasts as long as nothing else interacts with the system. If there is even the slightest interaction—one collision with a single molecule of air, for example—the system "decoheres" and the entanglement is lost. That's why quantum teleportation systems must take great pains to isolate their entangled pairs. Our favorite temporal physicist, Gene, incorrectly says that one can't separate an entangled pair: "You probably shouldn't even try." This is probably wishful thinking on his part. He associates entanglement with his doomed romantic relationship. But in fact, disentanglement is incredibly easy. It takes a great deal of effort to keep two particles entangled, just as Gene must construct an elaborate, time-stopping experimental setup to keep his girlfriend from breaking up with him in "Happy Anniversary" (A-2). Yet he is trying to save something that no longer exists. Once she decides it's no longer worth the effort, their "entanglement" is lost. Deep down, he senses the severed connection between them.

Initially Fred is as enthralled as the rest of Los Angeles by Jasmine's hypnotic power. But when she is exposed to Jasmine's blood in "Shiny Happy People" (A-4), the exposure causes Fred to "disentangle" from rest of her fellow "particles." The outside interference makes her decohere. By exposing each member of Team Angel in turn to Jasmine's blood, she ensures that they are no longer caught in that coherent state, and thereby breaks Jasmine's hold on them ("The Magic Bullet," A-4). Without coherence, entanglement is impossible. Decoherence is the reason the mass entanglement of subatomic particles doesn't seem to occur in our universe.

Or does it? In his book *Hyperspace* the physicist Michio Kaku speculates on the possibility of a cosmic entanglement between every atom in a person's body and atoms that are light-years away at the farthest reaches of the universe. He reasons that if the universe began as a quantum singularity, then the wave functions of all the particles in the big bang were once entangled, because they arose from a single mother state. So it's possible that their wave functions might still be partially connected, billions of years later, and that disturbances in one part of the Great Cosmic Wave Function can influence another, distant part.

Willow, for one, seems to adhere to this view, telling Giles that everything is connected, from the root systems of plants to individual molecules and energy ("Lessons," B-7). Skip, the seemingly friendly demon serving as Cordelia's guide on the astral plane in "Birthday" (A-3), echoes that sentiment when he tells her, "Inside every living thing there is a connection to the Powers That Be." Not to be outdone, evil has its own cosmic connection. A newly minted vampire in "Conversations with Dead People" (B-7) tells Buffy that he feels as though he is "connected to a powerful, all-consuming evil that's gonna suck the world into a fiery oblivion." From his demonic perspective, that's a good thing.

The notion of a Mother Wave Function is a bit disturbing, since if all things originate from a single quantum state, then everything that happens in the universe was imprinted at its birth. All outcomes were long ago established in the first few seconds of the big bang, and the cosmos has been evolving ever since according to a preordained pattern. Instead of human measurement and observation determining reality, our choices and actions become meaningless, because free will does not exist. The good news for those who believe in free will is that, while the universe might have started out as a coherent quantum system, any entanglement should have long since disappeared because of decoherence.

FELINE FATALE

Curiosity may have killed at least one cat in the Buffyverse. Early on in their relationship, Willow and Tara adopt a kitten and name her Miss Kitty Fantastico. The kitten makes a couple of subsequent cameo appearances, but eventually disappears without explanation—at least until "End of Days" (B-7), when Dawn makes an offhand reference to a household crossbow accident involving Miss Kitty, implying that the kitten was killed. Yet we, the viewers, have no way of verifying whether Miss Kitty is alive or dead unless we observe her state directly. This is the essence of Schrödinger's Cat, the eponymous thought experiment proposed by Schrödinger in 1935 to demonstrate the sometimes maddening, illogical absurdity of the subatomic world.

Perhaps Miss Kitty's encounter with the crossbow wasn't an accident. We only have Dawn's word to that effect, after all. Instead, imagine that Dawn sets up a fiendish experiment, confining Miss Kitty Fantastico to the corner of her bedroom, in front of a loaded crossbow whose trigger mechanism is connected to a

Geiger counter, and closes the door. Also present in the room is a single uranium atom (a highly unstable element) in a petri dish, right next to the Geiger counter. The uranium atom has a 50 percent probability of decaying and emitting an electron. If this occurs, the radiation will set off the Geiger counter, the crossbow's trigger mechanism will be released, and Miss Kitty will be killed by the sharp stake ejected from the crossbow. If the atom doesn't decay, then the trigger will not be released, and the cat will live. To find out what happened to the cat, we have to open the bedroom door and look inside. But what is the state of the cat before we do so? According to the strictest interpretation of quantum mechanics, poor Miss Kitty exists in two superimposed potential states: both dead and alive at the same time. It is the act of observation that decides her fate.

Stephen Hawking likes to joke that whenever he hears about the "cat problem," he reaches for his gun. A cat cannot be both dead and alive, not even in the Buffyverse—although it can be Undead, like the zombie cat in "Dead Man's Party" (B-3). Everything in our daily experience tells us that such a thing isn't possible. That was Schrödinger's point. Quantum mechanics only works for objects that exist at the subatomic scale; extrapolating those same principles to encompass the macroscopic world leads to absurd conclusions. Einstein once asked Niels Bohr—another early pioneer in quantum mechanics—whether Bohr truly believed that the moon is not really there when we don't happen to be looking at it. To Einstein, the entire notion of observation or measurement dictating experimental outcomes was ridiculous, since it denied the very existence of a solid, underlying reality. Even Schrödinger, inventor of the wave function, was deeply disturbed by the implications of what he'd helped create, memorably declaring, "I don't like it, and I'm sorry I ever had anything to do with it."

So why don't we see these weird quantum effects beyond the

$\Delta x \Delta p_x \geq \frac{1}{2}\hbar$?

subatomic world? Like everything else in quantum mechanics, it's primarily a question of large numbers. Large objects like Miss Kitty or the moon are made up of billions and billions of subatomic particles. These particles don't exist in isolation; they interact with other particles in their surrounding environment. That makes the wave function calculations for large objects extremely complex and difficult to solve. It's not enough to simply say that a ball rolled down the hill along a certain path. There are many different paths the ball could have taken to reach the exact same outcome. Out of all the possible paths, why did the ball "choose" that particular path? If just one or two subatomic particles are involved, the number of possible paths is limited. But when you're dealing with a multitude of particles, there are many different ways for them to interact, each with the exact same outcome. Quantum mechanics must consider every single option.

We can apply this concept to Dawn's fatal feline experiment. When the stake leaves the crossbow (assuming the uranium has decayed, releasing the trigger mechanism), it flies across a given distance and strikes Miss Kitty in the heart, killing her instantly. Dawn can determine how long it takes the stake to travel to its target, then use that information to calculate the exact path it took. That's simple Newtonian physics. If Dawn knows the position and velocity of any given object at a particular moment, she will be able to predict its position and velocity at any other moment, past and future. But if the stake is the size of a photon, the rules of quantum mechanics hold sway. Uncertainty dictates that she *can't* know both the velocity and final position of the stake. So Dawn's calculations would have to take into account all possible paths that the stake could conceivably travel before reaching its destination.

Probability still reigns supreme. Even though all possible trajectories contribute to the motion of the stake in flight, at the macroscale, the path predicted by Newton's laws has a higher probability and therefore contributes much more to the total sum than all the other paths combined. That's why those are the ones we see exclusively in our everyday world. But at the subatomic level, many different paths contribute at roughly the same level, and thus all play an equally important role in the averaging process, making quantum behavior much more difficult to predict.

The ability of superimposed potential quantum states to coexist is dependent on coherence. It's the only way Miss Kitty can be both alive and dead at the same time. Theoretically, before Dawn opens the bedroom door, there is one state where Miss Kitty is alive, and another where she is dead. These two states can coexist only if their respective wave functions are perfectly synchronized. They are coherent, and behave as a single quantum object.

But there's a problem when we take Schrödinger's abstract thought experiment into physical reality. Just as with entanglement, the slightest interaction with the outside world causes the wave functions of Miss Kitty's superimposed states to gradually fall out of synch, or decohere. The outside interference constitutes an act of measurement. Even before Dawn opens the door to see what happened to Miss Kitty, the bedroom environment has already completed billions of "observations." The wave functions collapse into a single quantum state: Miss Kitty is either dead or alive, not both at the same time. In 2005, NIST scientists managed to make half a dozen beryllium atoms inhabit a superimposed "cat state": The atoms were each spinning clockwise and counterclockwise at the same time, in perfect coherent synchrony. Yet once an atom "realized" the absurdity of its behavior and began spinning just one way, that coherence was broken. All the other atoms followed suit and the "cat state" vanished.

The upshot is that quantum physics appears to solve its own paradoxes. Decoherence is like a built-in fail-safe mechanism, ensuring that a large object made of billions of subatomic particles rarely behaves in a truly coherent fashion. It is extremely difficult to get more than a few atoms to vibrate together, perfectly synchronized, because of interference. In the real world, objects interact constantly with the environment, and decoherence occurs instantaneously. So Schrödinger's macroscopic-yet-quantum cat is an impossible beast.

The same basic principle applies to Marcy, the invisible girl, and to every other instance of large-scale quantum weirdness that occurs in the Buffyverse, including Jasmine's mass entanglement of Los Angeles. Even if her teachers and fellow students don't notice Marcy, the particles that make up her body are constantly interacting with other particles in her environment, a form of observation that would normally ensure that she is always in a decoherent state. It takes a great deal of effort and painstaking labor—not to mention a large and complex experimental setup—to maintain coherence in just the handful of atoms in a BEC, or the two gas clouds in the Danish entanglement experiment, much less the trillions of atoms in a human body, because of all the potential outside interference. So we can conclude that an even greater amount of effort would be required to maintain a coherent Marcy-state.

Where does the energy to fuel that additional effort come from? Jasmine derives her power to maintain coherence from feeding off her unsuspecting followers, killing thousands, she reasons, in order to save billions. In Marcy's case, extra energy is supplied mystically by the Hellmouth. The Hellmouth energy "makes it real," according to Giles. Somehow it keeps the atoms that make up Marcy in such a highly coherent state that quantum effects can manifest physically in the macroscopic "Marcy system." She becomes invisible.* The real world lacks such an infinitely powerful

*Technically, once her antics attract attention, Marcy should become visible again, particularly given her physical interactions with the targets of her animosity. But then there would be no plot.

energy source. This is a good thing; otherwise we'd be tripping over dead/alive cats, dodging our own manifested nightmares, and bumping into invisible high school students on a daily basis.

Decoherence also answers Einstein's question about whether the moon truly exists when we aren't looking at it. The moon does not exist in isolation. It interacts with everything around it, including the Sun. The rain of photons from the sun's rays onto the moon's surface constitutes a "measurement": The photons interact with the particles that make up the moon, collapsing their respective wave functions and causing decoherence. This gets rid of any superimposed states, with no need for conscious human interaction. So yes, the moon exists, independent of our perception. As physicist-turned-science–writer David Lindley so aptly puts it in *Where Does the Weirdness Go?*, "The world doesn't need us to look at it." If only the same could be said of poor Marcy.

7

ALTERNATE REALITY BITES

A Romp Through the Many Worlds of the Multiverse

"There are worlds infinite in number and different in size."
—DEMOCRITUS

Hell hath no fury like a woman scorned, particularly if she has a demonic ally. A heartbroken Cordelia naively makes a wish at the goading of a vengeance demon (Anyanka) that Buffy had never come to Sunnydale in "The Wish" (B-3). Her seemingly innocuous utterance brings about a horrific alternate reality—the embodiment of the adage "Be careful what you wish for, for you may get it." Only one small factor has been changed, and yet it changes everything.

"Time forks perpetually toward innumerable futures," the Argentine novelist Jorge Luis Borges writes in his short story "The Garden of Forking Paths." In the Buffyverse, some of those innumerable futures take physical form as alternate realities. Jonathan casts a spell in "Superstar" (B-4) creating an alternate reality where he excels at everything. On his wedding day, Xander is shown a nightmarish alternate reality of his future married life ("Hell's Bells," B-6). And Cordelia—who clearly didn't learn her lesson—gets to experience her ideal alternate reality in "Birthday" (A-3), where she is a famous TV star instead of a struggling actress who demon-hunts by night.

The notion that there is an infinite number of possible futures, or branches in the chronological tree, isn't confined to the Buffyverse, or Latin American literature. A handful of theoretical physicists has postulated the existence of other worlds, in which every possible forking of every possible path is ultimately realized in its own separate universe. Dubbed the "Many Worlds" hypothesis, it dangles precariously on the fringe of mainstream physics. But it serves as a handy mechanism to explain the possible origin of alternate realities in the Buffyverse.

ONE WORLD IS NOT ENOUGH

The far-reaching consequences of Cordelia's wish are an indication of just how critical a role Buffy has played in shaping the town's future. Shortly after arriving in Sunnydale, Buffy prevents the Master from rising, and rescues Willow and Xander from becoming snack food for the ancient bloodsucker. In fact, thanks to her diligent slayage, her graduating class boasts the lowest mortality rate of any in the high school's bloodied history. But if Buffy never arrives, a different progression of events must unfold, and does so, in gruesome detail. The Master rises because Buffy isn't there to stop him, and Willow and Xander are sired as vampires. Cordelia doesn't fare much better: She is brutally killed, and her body is cremated in the high school incinerator.

This was hardly Cordelia's hoped-for outcome, but vengeance demons can't be trusted. If only Cordelia had chosen to focus her rage on Xander (arguably a more deserving target), she might have found herself transported to a different alternate reality—preferably one in which she gets to live. After all, if one such reality exists, why shouldn't there be countless others, one for every possible version of the wish she makes? That's the philosophical underpinning for the Many Worlds hypothesis.

The concept as it pertained to physics first emerged in the 1950s, springing from the fertile imagination of a nuclear physicist named Hugh Everett III. Like many of his scientific colleagues, Everett was deeply dissatisfied with the troubling implications of quantum mechanics: specifically, the idea that observation determines reality. Recall from the previous chapter that in any quantum system (e.g., a subatomic particle like a photon or an electron), every possible outcome for an experiment is present simultaneously in a sort of superimposed limbo state. The sum of all those outcomes is described by an equation known as the wave function. It's only when we check to see what happened—that is, when we observe the system by making a measurement—that the wave function collapses and all those possibilities reduce to a single "real" event: the outcome we have observed.

But what happens to those other possibilities once the wave function has collapsed? The strictest interpretation of quantum theory has no good answer for this, and simply assumes that all the other potential outcomes vanish by necessity once a measurement is made. Everett came up with his own ingenious solution to the conundrum. Perhaps, he reasoned, a good wave function never dies; it just seems to collapse. Instead, it continues to evolve, forever splitting into other wave functions in a never-ending tree, with every branch becoming an entire universe. In this way, every potential outcome contained in the wave function—a photon appearing as a particle or wave, or Buffy coming (or not coming) to Sunnydale—is realized in its own separate universe.

Everett's hypothesis offers an alternative to resolving the nagging paradox of Schrödinger's Cat, or the uncertain fate of Miss Kitty Fantastico, who may or may not have been killed by a crossbow. Traditional quantum mechanics dictates that Miss Kitty (however briefly) inhabits two superimposed states—dead *and* alive—until some kind of observation takes place, either by Dawn checking to see what happened, or via decoherence: the interactions between Miss Kitty's particles and the billions of other particles in the atmosphere that surrounds her. But if time forks into two separate paths, forming two distinct universes, then there is no superimposed paradox. Instead of Miss Kitty being dead and alive at the same time, the fatal accident occurs only in one universe. In the other, Miss Kitty lives.

Worlds split when irreversible events occur, such as a choice, or in Cordelia's case, a wish. According to Everett, any measurement constitutes a kind of choice, because it forces each possible outcome to inhabit its very own universe. In one world, a photon will appear as a particle. In the other, it will appear as a wave. In much the same way, every choice (or action) we make alters the course of events. If we go back in time and change one of those choices, the result would be a separate universe where everything is identical up until that crucial decision is made. It's the act of choosing that makes one of those options "our" reality, but Everett insisted that all other options—the paths not taken—still exist somewhere, in a parallel universe beyond our ken.

Buffy is infected by the toxic venom of a Glarghk Ghul

Kashma'nik demon in "Normal Again" (B-6). The poison triggers sudden shifts between her life in Sunnydale and an alternate reality where she is a patient in a Los Angeles mental institution and the entire Buffyverse is a mere figment of her imagination. Buffy confesses to Willow that, before moving to Sunnydale, she was indeed committed to a mental institution by her parents after seeing her first vampire. Nobody believed her, so she chose to keep quiet about the monsters and her mission, until she was duly released.

Or was she? The Scoobies assume that Buffy is merely hallucinating, but it soon becomes clear from the episode's context that the alternate reality is just as real as Sunnydale. In many ways, it is eminently more sensible. There is no magic, no monsters, and no Chosen One fighting a lonely battle against impossible odds—just one delusional teenage girl. That moment represents an irreversible event, a pivotal fork in the road where her chronological time line diverges along separate paths. Buffy's world literally splits into two distinct physical realities. Everything up to that point in the time line is identical, but on one path, Buffy is released and moves to Sunnydale, where the "history" of the series unfolds. On the other, she never leaves the hospital, sinking so deeply into the fantasy world she has created that she is largely catatonic.

In a simple system, there are only two possibilities, giving rise to only two alternate universes. Miss Kitty is alive or dead; Buffy leaves the mental institution or she doesn't, which in turn determines whether Buffy comes to Sunnydale or not. But our choices in life are rarely so simple. Even if we start with two either/or options, the possibilities increase exponentially as the time line evolves. For instance, if Buffy is released from the mental hospital, moving to Sunnydale is not the only potential outcome. Her mother could decide to stay in Los Angeles or move to Pasadena—or even, God forbid, to Cleveland, which is rumored to have its very own Hellmouth. There are branchings upon branchings of possible choices, an infinite number of paths our lives can take, and each must give rise to its own alternate reality. So there must be an infinite number of other worlds, linked through a vast network of forks in the proverbial road.

Therein lies one of the primary criticisms of the Many Worlds approach: Taken down to the subatomic level, the notion quickly sinks under the weight of too many universes. The quantum world deals with incredibly large numbers. We've seen that determining the trajectory of a quantum-size stake fired from a subatomic crossbow requires taking into account all possible paths it can travel from its release to when it hits a target. Under Everett's theory, each path must have its own split-off universe. In the same way, a separate universe is needed even for the smallest, most inconsequential quantum event. In fact, the universe must split into roughly 10^{100} copies every second, all of them supposedly equally real—which means there must be billions upon billions of parallel worlds.

This is anathema to most physicists, for whom reductionism is a sacred tenet. The objective of physics is to reduce observable phenomena to the simplest, most economical explanation—ideally, to a mathematical formula as elegantly compact as $E = mc^2$. The debate between mainstream quantum physicists and proponents of Many Worlds echoes one of the oldest philosophical arguments in human history, dating back to ancient Greece. Plato envisioned a reality that existed in its own separate realm of ideal forms; all objects in our world are mere shadows, lesser copies of those ideals. The Platonic argument persisted well into the Middle Ages, when a medieval philosopher named William of Ockham pointed out that it is never desirable to multiply physical entities beyond what is strictly necessary.

"Ockham's Razor" supposedly shaved off Plato's beard, much like the reductionism of mainstream physics should shave away Everett's infinite number of parallel universes. The simplest, most economical explanation, scientists have found, usually proves correct. But the select few who still champion Many Worlds insist that their approach *is* the simplest and most economical explanation. It eliminates the need for superimposed states, and answers the question of what happens to the potential outcomes that weren't "chosen" when the wave function collapsed. It dispenses with the nagging question of how, exactly, nature chooses a particular path, because all choices are realized. And instead of having two sets of competing theories—relativity and quantum mechanics—applying to the macroscale and the subatomic world, respectively, Many Worlds applies the same quantum laws to large-scale living observers as it does to inanimate subatomic particles. Alas, the price is too many universes.

In the Buffyverse, as in quantum mechanics, reductionism takes the form of choice. In "Normal Again" (B-6), Buffy must choose whether to drink the antidote to the demon's venom and remain in Sunnydale with her friends, or not to drink and allow the demon to kill her beloved Scoobies (who are merely figments of her imagination in the mental-hospital reality). She can then inhabit the alternate reality, where her parents never divorced, her mother is still alive, and all the scary things that go bump in the night are nonexistent. Buffy ultimately decides to drink the antidote and rescue her friends, but either choice will serve the same purpose as collapsing the wave function: It will reduce her two conflicting worlds to one. And once Giles smashes Anyanka's power center (her amulet), Cordelia's wish is undone and the original time line of the series is restored. It's another example of a wave-function-collapsing, irreversible event. Sunnydale is now once again a single reality—at least as far as its inhabitants are concerned. But perhaps the wave function only seems to collapse. If Many Worlds holds sway in the Buffyverse, those other alternate realities still exist somewhere, along with countless others.

SEEING DOUBLE

Without her power center, Anyanka is rendered mortal, and becomes just Anya. She is also the only one who remembers the alternate reality. In "Doppelgangland" (B-3), Anya tricks Willow into performing a spell to create a temporal fold. Her intent is to go back in time to the alternate reality and retrieve her amulet, thus restoring herself as a vengeance demon. But there is a glitch. Instead of the amulet, Willow's vampire self is transported from that alternate world into the original Sunnydale reality.

This causes all kinds of confusion and mayhem, as Vamp-Willow enlists her very own gang of vampires to re-create the nightmarish version of Sunnydale that she has lost. It's not quite a case of mistaken identity, since both versions are the real Willow in their respective worlds. The two are genuine doppelgängers.

The existence of doppelgängers is inherent in Many Worlds, since if a separate universe exists for all possible outcomes of an experiment, each universe must contain not just duplicates of the equipment used to perform the experiment, but also "copies" of the scientist who performs it. However, the theory insists that once we perceive a given reality, we are stuck with it. We can only experience one reality at a time, at least until we develop the ability to fold back space-time as easily as Anya and Willow do, equipped with nothing but chicken feet, magic sand, and a whole lot of extra mystical energy. That's because any form of communication, interaction, or interference between alternate universes—defined as realities containing different outcomes of the same measurement—would violate quantum mechanics. The stipulation serves the same purpose as decoherence. We can't see both the particle and wave nature of a photon at the same time, because this awareness constitutes an observation, destroying the system's coherence and triggering the wave-function collapse. So only one result can be obtained for any given measurement in each universe.

This is where the Buffyverse strays from orthodox Many Worlds theory. There are two different versions of the same person in a single reality: the rather skanky, leather-clad Vamp-Willow, and her human Willow counterpart, distinguishable (to the viewer, at least) by her fuzzy pink sweater. Eventually the two Willows come face-to-face. Strictly speaking, that meeting should never take place, because one can never experience more than one reality at a time; they must be kept absolutely separate. Take away that safeguard, and an alternate reality creates uncomfortable paradoxes, like two versions of the same person in one reality.

If we could send and receive information from our world to an alternate reality, not only would we be aware of its existence, we could in turn have an effect on that other world. Some of these effects would be minor. Vamp-Willow's smack-down of Percy, the arrogant jock, results in Percy treating human Willow with considerably more respect thereafter. But other actions can have serious consequences.

When Cordelia is transported to the nightmare world of "The Wish" (B-3), she still remembers her former reality. This knowledge affects her actions, which in turn alter the sequence of events in the nightmare reality. Her presence creates yet another alternate reality, branching off from the first one. The divergence begins when Cordelia tells Giles that Buffy should be in Sunnydale—right before she dies. Giles tracks down this alternate Buffy, asking for her help. A much tougher and more cynical Buffy arrives belatedly in Sunnydale, attempts to kill the Master, and is killed by him instead. Before she dies, she stakes Vamp-Xander. Oz helps impale Vamp-Willow on a protruding piece of wood. Even Angel meets the pointy end of a stake, foiling any chance of getting his own spin-off series in that alternate reality. Would any of this have happened if Cordelia didn't remember her prior world? Maybe. Maybe not. Events certainly wouldn't unfold in quite the same way.

Sometimes the Buffyverse gets it exactly right. The history of the entire series was rewritten with the introduction of Buffy's younger sister, Dawn, in Season 5. Dawn turns out to be a mystical Key in human form. She is placed in the Summers household by an order of monks, who alter the memories and history of everyone associated with Buffy's family from the moment Dawn was supposedly "born." So the entire series from Season 5 onward becomes an alternate reality. Nobody remembers a time when Dawn wasn't in the picture, least of all Dawn herself. And the process is irreversible. Once worlds split, there is no going back. Even when the Scoobies and Buffy's mom, Joyce Summers, realize who (or what) Dawn is, they still have no memory of their lives without her. She remains part of their new reality.

Nonetheless, Everett thought that worlds might occasionally fuse, provided that all records, memories, and other information that distinguishes between the two fused worlds is lost. When Angel makes a deal with Wolfram & Hart to provide a happier alternate reality for his son, Connor, one of the requirements is that no one else remembers the other reality. In "Origins" (A-5), Wesley smashes the mystical orb that holds Connor's former memories. The two sets of memories merge, bringing the two forked paths of Connor's life together, after a fashion. Only the memories return; everything else about Connor's altered life is intact. It's not a perfect fusing as Everett envisioned it, but it's close enough for a fictive universe.

The dictum against communication or interaction between separate realities would explain why we are not aware of being split off every microsecond into countless multitudes of separate universes. But it also means that the Many Worlds interpretation can never be definitively tested or proven. If the separate universes are noninteracting, there is no possibility of doing an experiment in one universe that would reveal the existence of another. That's the other primary argument against Many Worlds. If a theory can't be tested, it's not really physics. Many Worlds might not be outright fiction, but mainstream physicists are inclined to consider the theory more akin to metaphysics—which might be why it fits so well within the context of the Buffyverse.

BRINGING UP BABY (UNIVERSES)

The Buffyverse contains more than alternate realities; it is also home to alternate dimensions.* Cordelia is accidentally transported to a strange new world in "Over the Rainbow" (A-2), and Angel, Wesley, and Gunn soon follow to rescue her. The world is called Pylea, and it is just one of a myriad of dimensions that populate the Buffyverse. There are hell dimensions, heavenly dimensions, the Oracle's timeless dimension, and mystically maintained prison dimensions, like the ones Wolfram & Hart operate for recalcitrant employees ("Underneath," A-5). Ghost-Spike gets caught in a spectral dimension in "Hellbound" (A-5), while Angel's son, Connor, is kidnapped shortly after his birth and grows up in the

*In the Buffyverse, "dimension" usually denotes a separate world or parallel universe, not a description of physical space—although there are occasional exceptions.

Quor-Toth dimension. To find the key to defeating Jasmine, Angel must travel to a dimension with an atmosphere that is toxic to humans, very much like that of another planet. There is even mention of a Land of Perpetual Wednesdays ("Triangle," B-5), as well as a World with Nothing but Shrimp ("Superstar," B-4).

Where did all these other dimensions come from? The shrimp dimension appears to have been created as a cosmic joke when Anya explains the concept of alternate realities to Buffy in "Superstar," but most of the other dimensions don't seem to be related to alternate realities at all. An alternate reality is a "copy" universe that has split off from another, identical to its parent in all respects up to the point where the chronological path diverges. Buffy coming or not coming to Sunnydale creates two distinct versions of her universe, as does Anya's hypothetical wish for a world without shrimp ("or, you know, nothing but shrimp"). In contrast, Pylea is a completely separate place, with different physical characteristics. Such dimensions don't contain exact copies of the objects, people, and events of another reality. They are unique. In a few cases, the series seems to blur the distinction. For instance, in "Triangle" (B-5), Anya and Willow magically teleport a rampaging troll back to the Land of the Trolls. Anya refers to it as an alternate reality, but from the context, it more closely resembles a separate dimension.

Many Worlds doesn't apply quite so neatly when considering alternate dimensions like Pylea. Fortunately, Everett's hypothesis is not the only scientific theory that speculates on the possible existence of other universes. The answer to the question of origin may be hidden in the mysteries that still shroud the birth of our own universe.

Forget what we've learned about mass/energy conservation for a moment. The laws of physics inexplicably break down at the universe's moment of birth, much as they do at a black hole's singularity. Technically, our universe arose out of nothing. We owe our existence to a random fluctuation in the quantum vacuum. At the subatomic scale, space is not the smooth, flat geometric entity that Einstein envisioned when he devised relativity. It is frothing with virtual pairs of particles and antiparticles that pop into existence for the briefest of moments before annihilating each other. Antimatter is the mirror image of ordinary matter, made up of antiparticles instead of ordinary particles. Their masses are the same, but their electrical charges are reversed. Virtual pairs include one of each, which is why they annihilate each other and convert back into pure energy.

The nascent universe was very dense and incredibly hot—tens of billions of degrees—so energy and mass were entirely interchangeable. New particles and antiparticles were constantly being created and colliding with their polar opposites, thereby annihilating back into energy. At some point in those first few moments, a small surplus of matter appeared. The imbalance was sufficient to wipe out all of the antimatter in the universe in about one second. As the universe expanded, the temperature dropped until it was too low to create new particle and antiparticle pairs. Only a small amount of "leftover" particles of matter remained, which make up every observable object in the universe we see today.

Initially, our universe was just a tiny bubble floating in a vast empty sea of space-time. But in the first trillionth of a trillionth of a second of its life, a mysterious anti-gravity force (negative energy) caused the universe to expand rapidly in a very short period of time. Scientists know that this cosmic growth spurt occurred, because they have measured minute temperature fluctuations in the cosmic background radiation—the remnant of all those particle/antiparticle collisions—that serve as a relic of inflation. Michio Kaku likens these fluctuations to seeds that expanded enormously as the universe cooled and spread outward, blossoming into the galactic clusters and galaxies we see in the universe today.

But why should inflation have only happened once? Scientists still don't know what sort of mechanism caused inflation, so they can't say with certainty that it couldn't happen again. Stephen Hawking has proposed a new twist that treats the entire universe as a single quantum particle. It's not as far-fetched as one might think. After all, at one point the entire universe was smaller than an electron, so quantum principles would have applied. In quantum theory, every particle is associated with a wave, which in turn tells the probability of finding the particle at any given point by considering its every possible state (the wave function). If the universe was once the size of a subatomic particle, it also must have a wave function. And that wave function must, by definition, describe not just the universe we know and love, but all possible universes. So our universe is not necessarily the only one in existence—just the most likely one.

Several cosmologists have speculated that at the Planck scale—the smallest possible distance, on a par with the smallest virus—not just particles, but tiny wormholes may spontaneously spring into existence, some of which may create a temporary "bridge" from which a new universe can spring. A tiny, random patch of universe could suddenly inflate at any time, sprouting a baby universe, which may in turn bud another, and so on, creating an extensive, interlocking web that cosmologists have dubbed the "multiverse."

In this scenario, big bangs happen all the time. There could be an infinite number of baby universes out there—most of them so small that we wouldn't even notice them branching off. Most inflate just a little after birth before recollapsing into nothingness, but probability dictates that eventually one will experience a period of inflation that lasts long enough to create a full-size universe.

The notion opens up the hypothetical possibility that a demon-infested dimension like Pylea, not to mention the entire Buffyverse, *could* exist in its own separate universe. As with Many Worlds, this theory appears to be inherently untestable, since we should not be able to observe or interact with other worlds in the multiverse. But even Gene, the time-obsessed physicist, isn't averse to considering the possibility. When told of the existence of demons in "Happy Anniversary" (A-2), he takes the information in stride, remarking that "the idea of demons among us is consistent with [the] multiverse."

LAWS AND ORDER

One of the first things the members of Team Angel notice when they arrive in Pylea is that the laws of physics appear to be slightly different from those in their own world, particularly with respect to vampires. Angel doesn't spontaneously combust when hit with sunlight, and he revels in the novelty. ("Did everybody notice how much fire I'm not on?" he exults.) He can also see his reflection for the first time in more than 200 years. In that respect, the physics of Pylea appears to be much more consistent with the laws of real-world physics than those that govern the world of Sunnydale and *Angel*'s Los Angeles.

Pylea mostly exhibits superficial differences. For example, it has two suns. But whether we're talking about the alternate dimensions of the Buffyverse or parallel universes in the multiverse, it's possible that such places may contain very different physical laws, as well as entirely different sets of subatomic particles, atoms, chemical interactions, and cosmological constants. These variants would play a critical role in determining whether a nascent universe inflated and expanded to full size or simply popped out of existence in a nanosecond. The variants would also determine the amount of matter present in the universe, as well as how it's distributed, and even whether such a universe would have conditions that are conducive to life. The multiverse is believed to contain billions of parallel universes, but most of these are devoid of life. However, the law of averages dictates that in at least one, a series of cosmic accidents would give rise to the physical characteristics needed to sustain life—and possibly more than one. Perhaps in one of these newly inflated universes, an entire race of horned, green-skinned demons evolved, or any other of the colorfully eclectic demon species that populate the Buffyverse.

Time itself could be altered in such universes. If the multiverse cosmology is correct, each separate universe within it would have its own self-contained time—an aspect that is entirely consistent with the Buffyverse. Time passes at different rates in each dimension. When Buffy must send Angel to hell in "Becoming" (B-2), he spends 100 years there from his perspective, but when he returns to Sunnydale, a mere few months have passed. A runaway Buffy rescues a group of street kids from a demon dimension in "Anne" (B-3), where they have been kidnapped and enslaved. Most toil for years and die of old age before anyone back in their home dimension realizes they're missing. And Connor is kidnapped into the Quor-Toth dimension as an infant, reappearing a few months later in "The Price" (A-3) as a rebellious eighteen-year-old intent on killing his father.

The same, no doubt, holds true for Pylea, although no direct comparison of the passage of time is explicitly stated. We are told that Fred has been missing from Los Angeles for five years, but not how long this is from her exiled perspective. All she says is that it felt like she had always been there, and that her former life was merely a lovely dream. Since she hasn't aged appreciably more quickly (or more slowly) than she would have back in Los Angeles, we can infer that the time differential is less extreme in Pylea than it is in Quor-Toth, for example.

Einstein's notion of time dilation offers a plausible explanation for the differences. Anya explains in "Triangle" that other dimensions are constantly pressing up against our world, and that their positions are always shifting. She compares teleporting objects between worlds to "trying to hit a puppy by throwing a bee at it"—a relativistic nightmare. Each dimension clearly constitutes its own frame of reference, since it is an entirely separate place. It is also moving with respect to all the others. So naturally time moves differently in each dimension, as it would in each universe of the multiverse. In certain cases, that movement would have to be significant. For Connor to go from infant to teenager in a mere few months, Quor-Toth would have to be moving more slowly than Angel's Los Angeles. And if Angel's hundred years in a hell dimension corresponds to three months in Sunnydale, the hell dimension would have to be moving significantly more slowly than Sunnydale.

Time is just the tip of the iceberg. If every material object in our universe is the result of a slight asymmetry between matter and antimatter, what would have happened if antimatter, instead of matter, had gained that initial advantage? Technically, this would create a mirror universe. Everything would be made of particles with identical masses, but with the charges reversed. Pylea seems to contain regular matter, so we can safely assume that a surplus of matter appeared shortly after at its birth, much like it did in our own universe.* Should there be any doubt, imagine

*If there were equal amounts of matter and antimatter, with neither side ever gaining the upper hand, there would be nothing left as the universe cooled,

what would happen if Team Angel traveled to an antimatter dimension. The minute they passed the portal threshold, the particles in their bodies would meet the antiparticles in the mirror universe and immediately annihilate into radiation. Since this doesn't happen, Pylea and *Angel*'s Los Angeles are both clearly material worlds—or both antimaterial worlds—but they are the same kind of world, not polar opposites.

Pylea is populated with living beings, so we can infer that its fundamental forces (strong force, weak force, gravity, and electromagnetism) are also quite similar to our own. Otherwise conditions would not have been conducive to life. The strong nuclear force holds atomic nuclei together. If it were just the slightest bit weaker, those nuclei would fly apart, and none of the elements found in today's universe could have developed in the interior of stars via nucleosynthesis. If it were a wee bit stronger, stars would burn their nuclear fuel too quickly and life could not evolve.

The same is true of the weak force, which controls interactions between atomic nuclei and neutrinos, the tiny particles created when neutrons become protons, or vice versa. The weak force only kicks in if neutrinos are so close as to be practically touching the atomic nuclei. When this happens, the atom turns into a different chemical element. If we vary the strength of the weak force even the tiniest bit, life becomes impossible. Neutrinos are needed to carry energy outward from an exploding supernova. This energy in turn is responsible for the creation of the heavier elements beyond iron. If the weak force were weaker, neutrinos would interact hardly at all, so nothing beyond iron could exist. If it were a bit stronger, neutrinos might not be able to escape from a star's core, again preventing the creation of the heavier elements.

---

thereby creating a vast, empty wasteland containing nothing but leftover radiation.

On a more cosmic scale, if gravity were too weak, stars could not condense and create the enormous temperatures necessary for fusion. The stars would not shine, and planets would plunge into freezing darkness, quashing any chance of life. Planets would not be able to form at all. If gravity were a bit stronger, the stars would heat up too rapidly and burn up their fuel too quickly to enable life to develop. Galaxies would also form earlier and would be much smaller than the ones we observe in our universe. The stars would be more densely packed.

In fact, Pylea might not have inflated into a full-size universe at all if conditions hadn't been exactly right at the moment it popped into the sea of space-time. The Greek letter *omega* (Ω) designates how densely matter is distributed throughout the universe—a property known as critical density—and this determines the strength of gravity. If *omega* were too small, the fledgling world of Pylea would have expanded and cooled too fast. If *omega* were too large, Baby Pylea would have collapsed before life could start. Fortunately for the residents of Pylea, in their case (as in ours), *omega* turned out to be just right.

The "cosmology" of the Buffyverse clearly contains elements of both Many Worlds and the multiverse, while not being entirely consistent with either theory—and arguably not even within itself at times. That's why it's science fiction. Nor is there any concrete scientific evidence to date for the existence of parallel universes in our own reality. But whether we're talking hypothetically about innumerable alternate realities, or an array of parallel worlds connected by wormholes into one gigantic multiverse, one predominant attribute remains the same: there is no such thing as an insignificant detail, either in the Buffyverse or in our own world. Change one small thing, and everything changes—right down to our very existence.

WEDNESDAY

8

ANY PORTAL IN A STORM

How to Build a Big, Swirly Wormhole

"I must warn you, we're going to pass through, well, a sort of gateway thing. It may disturb you. It scares the willies out of me."
—Slartibarfast, THE HITCHIKER'S GUIDE TO THE GALAXY

In "She" (A-1), a comely female demon named Jhiera arrives in Los Angeles by means of a mystical portal from her home dimension of Oden-Tao. In her world, the women are enslaved. Jhiera has escaped and is essentially running an interdimensional underground railroad to help other females from her world do likewise, with a posse of the males of the species in hot pursuit.

Thanks to the portal mechanism, Oden-Tao's war between the sexes spills into an entirely new dimension.

Cordelia is understandably dismayed to learn that dimensional portals exist. The members of Team Angel have enough trouble fighting the demons in their own dimension, never mind the occasional imported creatures from another. Her dismay would be shared by many scientists, since they have yet to uncover any direct evidence that portal-like tears in the fabric of space-time can even occur, much less be used to transport large living objects from one world to another (assuming other worlds exist). The skeptics would include adherents to Many Worlds, since that theory insists on separate universes being "causally isolated" from each other. They cannot interact with other universes, and must be entirely self-contained.

The notion of a multiverse is essentially a reworking of Many Worlds, with one critical difference. In the multiverse, the dictum against interactions isn't absolute. Travel between worlds is technically within the laws of physics, but statistically highly unlikely. One might be inclined to think that baby universes should be self-contained and noninteracting. However, once one postulates the existence of wormholes, it raises the possibility of using those hypothetical structures to travel between linked worlds, although the practical difficulties associated with making such structures traversable are pretty intractable. Just the probabilities of a wormhole spontaneously opening in our immediate vicinity are so minuscule that we would have to wait a very long time—longer than the entire lifetime of our universe—before such an event occurred.

That probability is much greater in the Buffyverse. In fact, the phenomenon happens quite frequently, even though we are told explicitly in "Spiral" (B-5) that the various dimensions are meant to be kept separate from one another. Portals nonetheless exist between them—Sunnydale's Hellmouth is one example—just like the wormholes that may link baby universes. Characters will insist now and then on opening a portal, even though the end result of doing so is almost never positive.

A BRIDGE TOO FAR

On an otherwise typical night at Caritas, the demon karaoke bar owned by Lorne, a mysterious, big swirly hole opens up onstage and spits out a hell-beast ("Belonging," A-2). The vortex is a portal connecting Los Angeles with Lorne's home dimension of Pylea, and the hell-beast is called a drokken. Lorne enlists Team Angel's help to track and kill the drokken before it feeds on too many local residents. So begins an adventure so big that it spans the next three episodes. Angel and his cohorts discover more than Cordelia ever wanted to know about dimensional portals, even traveling to Pylea themselves.

One of the Pyleancentric episodes is called, aptly enough, "Through the Looking Glass" (A-2), a reference to Lewis Carroll's nineteenth-century sequel to *Alice in Wonderland*. Carroll's looking glass resembles a wormhole that connects Alice's world with a strange new land, and wormholes are the closest concept in real-world physics to the mystical portals found in the Buffyverse. Technically, a wormhole is defined as any structure that connects two otherwise distant or unrelated points in spacetime. The notion is an outgrowth of black hole physics. We've seen that a black hole has at its center a point of infinite density and infinite gravity, known as a singularity, as well as a theoretical point of no return called the event horizon beyond which nothing escapes—not even light.

Einstein thought that black holes were merely theoretical objects, but mathematically, he showed that there might be a wormhole at the heart of one. According to the equations of general relativity, wormholes can arise when black holes bend space-time far enough to poke a hole through to another point in space-time. Picture a bedsheet stretched taut. That sheet is Einstein's flat, geometric space-time. Place a large bowling ball in the center of the sheet, and the sheet will bend inward in response, creating a gravitational pull. Now imagine that the bowling ball is being squeezed, so that the same amount of mass must fit into a smaller and smaller space. The ball will become denser and denser as it becomes smaller and smaller, causing the sheet to dip lower and lower, until finally the ball has been squeezed down to the size of a (very heavy) pinhead. At that point, its density becomes so great, and the resulting gravitational force so strong, that it pokes a small hole in the center of the sheet.

That's exactly what would happen if a wormhole formed at the center of a black hole: For a brief moment, there is a tear in the fabric of space-time. But what lies on the other side? Always a stickler for symmetry in his equations, Einstein hypothesized that a "mirror universe" must exist on the other side. A black hole can be visualized geometrically as a large funnel with a long throat. If one "cuts" the throat and merges it with a second black hole that has been flipped over (a "white hole"), the resulting configuration resembles an hourglass, with the two ends connected by a thin filament. This so-called Einstein-Rosen bridge (named for Einstein and his collaborator, Nathan Rosen) is an early theoretical incarnation of a wormhole, serving as a back door leading from the interior of one black hole into another.

Mathematicians call such hypothetical formations "multiply

connected spaces." To get a sense of how they work, consider a simpler (mathematically speaking) model for a wormhole called Misner space, which is the brainchild of Charles Misner, a theoretical physicist at the University of Maryland, College Park. Imagine that Buffy's bedroom comprises the entire universe, and the walls correspond to the mouths of a wormhole. For instance, every point on the wall in front of her is identical to the corresponding point on the wall behind her. The two bedroom walls are joined, like a piece of paper that is folded over so that its two ends meet, forming a cylinder. So if Buffy walks through the front wall, she will reappear from the back wall.

It's a bit like how old video-game screens used to wrap the field of play around both the left and right edges of the computer monitor. In fact, if the bedroom is small enough—barely able to contain Buffy's body—and she passes her hand through the wall in front of her, she could reach completely around her little universe to grab her own shoulder when her hand reemerges from the wall behind her. In much the same way, the mouths of a wormhole create a shortcut between two points in space. If each of those points is a separate universe, rather than different points in Buffy's bedroom, the wormhole bridges two worlds.

Wolfram & Hart's torturous prison dimension in "Underneath" (A-5)—where the senior partners confine renegade lawyer Lindsey McDonald—resembles a Misner space. On the surface, it appears to be a pleasant, middle-class suburban street with rows of identical houses stretching into the horizon. (The horrific rituals that go on in the basements of those houses are far from idyllic, however.) But if Lindsey tried to leave the dimension by walking down that street toward the horizon, eventually he would hit a "wall" and find himself right back at the other end of the street. In this case, the two "mouths" of the Misner "wormhole" connect opposite ends of the prison dimension, rather than lead-

ing to another world. The only way out is through a separate, interdimensional portal—a fiery opening in the dreaded basement, aptly called "the Wrath."

As recently as 1990, most scientists agreed with Einstein that black holes were theoretical entities—an intriguing mathematical anomaly, to be sure, but largely the stuff of science fiction. However, several hundred such objects have since been identified, and as many as 300 million might exist in the visible universe. Astronomers now believe that most of the galaxies in our universe have black holes at their centers; a black hole was identified at the center of our own Milky Way. But the existence of wormholes is still open to debate. Scientists have yet to uncover any hard, physical evidence of one. That's partly because we can't see inside a black hole to observe directly what lies at the singularity, and partly because wormholes would be extremely difficult to make. Enormous amounts of mass and/or energy are required to achieve the extreme curvature of space-time that would give rise to a macroscopic wormhole—energies far beyond our present capabilities. And as Wesley says in "Over the Rainbow" (A-2), "There's obviously not going to be any big, swirly hole-jumping without a big, swirly hole."

Fortunately, staggering amounts of extra mystical energy are readily available in the Buffyverse, provided one knows how to harness it—and, in the case of dimensional portals, where to look. The gang learns that portals only open in certain "hot spots," areas of concentrated psychic energy that serve as natural gateways to other worlds. Phrased in less mystical terms, one could say that the presence of this extra energy means that the fabric of space-time is more vulnerable to bending and tearing in those areas, so it can more easily rearrange itself to create the requisite portals when certain conditions are met.

The process is considerably easier in the Buffyverse than it is in our own universe. To open a portal to Pylea, one must read aloud from an ancient book written in a language comprised entirely of consonants. Mystical languages possess great power in the Buffyverse, and Fred discovers why this might be the case: The book in question doesn't contain actual "words," but consonant representations of mathematical formulae for transfiguring the fabric of space-time. Even without such a book, Fred's physics training enables her to calculate the requisite math for opening portals. She just doesn't know where the hot spots are located, and thus assumes that her equations are wrong. In fact, she unwittingly opens several portals while trapped in Pylea.

AS THE WORMHOLE TURNS

When Cordelia is accidentally sucked into Pylea, Angel decides to open another portal and go in after her. After all, how hard can it be to read aloud from a book, even if the text is all consonants? Alas, things are never that simple and straightforward in the Buffyverse. There is always a catch. For instance, in keeping with the laws of thermodynamics, psychic hot spots often "go cold" after use. The hot spot located in Caritas is drained of its energy when Cordelia is sucked through, but Angel can't wait for it to recharge. He must find a second hot spot in order to open another portal.

Furthermore, whenever individual entities enter a dimensional portal, they tend to separate in transit. So the remaining members of Team Angel wouldn't arrive together, even if they jumped through the portal at the same place and time. They could even end up on opposite ends of Pylea. Wesley considers performing a binding spell to prevent such a separation, but there's a catch there too. While passing through the portal, the effect would be magnified to such an extent that Angel, Wesley, Gunn, and Lorne could end up fused together like some bizarre four-man (or three-man, one-demon) Siamese quadruplet.

The problems are different in real-world physics, but the situation isn't any less complicated. Let's assume that wormholes really do spring spontaneously from the foamy fabric of spacetime, with no need for reciting from mystical books written in strange consonant languages. That doesn't make them traversable. Such wormholes would be tiny, about 100 billion billion times smaller than a proton at the center of a single atom. This is far too minuscule to allow for human travel. And most of those tiny wormholes would be unstable. They would only exist for fractions of a second before recollapsing into the quantum foam. Light is the fastest entity in the known universe, yet even a beam of light traveling at top speed wouldn't have time to pass through a tiny wormhole. How can we inflate a wormhole to macroscopic size and then hold it open long enough for a space probe to pass through to the other side?

That's not the only problem. Remember the sad (hypothetical) fate of Gene, the time-obsessed physicist, should he happen to fall into a black hole? Once Gene crossed the event horizon, he would be sucked inexorably toward the center and be subjected to such intense gravitational fields that he would be torn apart. Even if he could reach the supposed wormhole at the center in one piece, he would be crushed to death by infinite gravity as he attempted to pass through to the other side—a much more dire effect than simply fusing Team Angel into a freakish Siamese quadruplet. Wesley solves Team Angel's dilemma when he discovers that iron or other metals counter the effect. By making the jump in Angel's convertible, they can all arrive together (and unfused). This is completely unrealistic from a physics standpoint, although the sight of the convertible flying through the big swirly hole certainly makes for fantastic television. An even more exotic solution is required in the world of physics.

Physicists thought that they might have found this solution when they discovered that all black holes observed thus far appear to be rotating, some as fast as a million miles per hour. This changes the scenario significantly, because it gives a black hole angular momentum. Conceivably, such an object could produce a stable wormhole. Recall from chapter 4 that Buffy whirls an incense burner on a chain around her head and then releases it to knock Spike out. The rotation creates a centripetal force that pushes outward, much like a merry-go-round. A similar phenomenon occurs with a rotating black hole.

A rotating black hole has two event horizons: the outer horizon that marks the point of no return for objects falling into the black hole, and an inner horizon closer to the center, arising from the hole's angular momentum. This type of black hole would remain stable because of the intense centripetal force pushing outward, canceling the inward pull of gravity and holding the portal open long enough to allow an object safe passage. Instead of collapsing down into a singularity, a rotating black hole would collapse into a ring. If an object approached this ring gradually, from the side, spiraling closer and closer with each pass, it would still encounter infinite gravity and space-time curvature and be destroyed. However, if the object managed to travel straight through the ring, the gravitational force would be large, but not infinite. It would have a shot at survival. It's a bit like how the water in a flushing toilet spins around the center of the bowl, whirling faster and faster before it is flushed through the hole at the bowl's center. If the members of Team Angel fell into a rotating black hole, they might not be crushed to death at the singularity if they fell directly into the ring. Instead, they would be "flushed" completely through the wormhole to the universe on the other side.

There is ample visual evidence that a similar rotating model describes many of the dimensional portals that exist in the Buffyverse. Portals to Pylea appear as big swirly holes. Buffy's first college roommate, Kathy—a demon who ran away from home to get a college education—is sucked back to her dimension through a swirling vortex when her father finally tracks her down in "Living Conditions" (B-4). Angel accesses a prison dimension operated by the Powers That Be via a mystical key and coin in "That Vision Thing" (A-2). Used in tandem, the objects fit together to form a spinning top that creates a swirling vortex. And in "Get It Done" (B-7), the Scoobies are watching Slayer Shadow Puppet Theater. Carved flat figures mounted on a spinning carousel tell the story of the first Slayer by casting shadows. The carousel starts spinning faster and faster, creating yet another vortex-like portal to another dimension.

Alas, even wormholes at the heart of rotating black holes are inherently unstable. Any accelerating object generates ripples, known as gravitational waves, in the fabric of space-time. This gravitational radiation travels at the speed of light, and could be amplified to infinite energy as it approached the black hole's singularity, warping space-time around itself so severely that it shuts the "door" to the other universe. The smallest disturbance could cause a wormhole to collapse, closing the entrance to the passage.

This might explain why portals in the Buffyverse only stay open for short periods of time and close abruptly once someone has jumped through them. The person's mass generates gravitational waves, which then amplify to such an extent that the portal collapses, or closes. It also means that any passage through a wormhole would be a one-way trip. Once you leave a universe, there is no turning back. Assuming a person wasn't crushed to death, he/she would need to open a second wormhole connecting the parallel universe back to ours in order to make the return trip. This is certainly how things work when it comes to Pylean portals. The members of Team Angel successfully make the jump into Pylea, but the book they use to open the portal in Los Angeles only works in that dimension. The book remains behind when they shoot through the portal. It doesn't even exist in Pylea. The members of Team Angel must use the priests' books to locate the relevant "hot spots" there before they can open up a new portal in order to travel back to their home dimension.

NATURE ABHORS A QUANTUM VACUUM

There is another possible solution to the collapse problem. Ordinary matter simply isn't strong enough to withstand the gravitational crush of space-time inside a black hole. Some form of antigravity (negative energy) that pushes outward instead of contracting inward would counteract a black hole's deadly squeeze, holding it open long enough to allow an object to pass through a hypothetical wormhole at its center. Since gravity arises from the curvature of space-time in response to the mass of a given object, the most likely source of antigravity would be an exotic form of negative matter. This is not the same thing as antimatter, which is comprised of antiparticles that have the same (positive) mass but opposite charges as their material counterparts.* Negative matter would have negative mass, producing negative curvature of space-time, giving rise to antigravity.

*Physicists have yet to find evidence that antimatter exists naturally in the universe today, but they have been able to create small amounts of antiparticles in atom smashers. Those antiparticles exist for mere fractions of a second before annihilating into radiation.

It's still very much in the realm of theory, but the idea is that perhaps a wormhole is not empty, but instead contains a "scaffolding" of negative matter that exerts an outward antigravitational push on its walls, sufficient to make the wormhole stable. The catch—and it's a big one—is that negative matter is purely hypothetical. Scientists have yet to observe even indirect evidence of its existence. In contrast, some form of negative energy is physically possible in very small amounts, but is extremely rare—and fleeting.

For better or worse, negative energy seems to be abundant in the Buffyverse. Illyria has the ability to rearrange the fabric of space-time at will to create dimensional portals whenever she needs one. What is the source of her power? We find a clue when Wesley saves Illyria from self-destructing by siphoning off a large portion of her demonic energy with a device he calls a "Mutari generator." This supposedly creates a pinhole through which Illyria's excess power can be channeled into an "infinite extradimensional space." He also describes it as a "negatively charged pocket universe."

Wesley might not be entirely sure of the exact nature of this mysterious location, but his comments hint that at least some of the excess mystical energy floating around the Buffyverse might be negative energy—the vital component for the creation of stable wormholes. If that's the case, no wonder dimensional portals keep opening up all over the place. It hardly seems fair. The real world lacks such tantalizing fictions as negatively charged pocket universes and Mutari generators. But we do have our very own potential source of negative energy in the fluctuations of the quantum vacuum.

The nineteenth-century Serbian inventor Nikola Tesla believed that the vacuum held enormous reservoirs of energy, sufficient to revolutionize human society if we could only devise a means of harnessing it. Tesla was highly eccentric—he ended his days impoverished and with questionable sanity—but his intuition about there being energy in the vacuum turned out to be true. We've seen that empty space isn't really empty. It roils and boils with quantum fluctuations, occasionally spitting out pairs of "virtual" elementary particles and antiparticles. These virtual particles and antiparticles annihilate and disappear back into the quantum vacuum so quickly that the apparent violation of energy conservation incurred by their creation can't be observed directly.

So how do we know they exist? There is indirect evidence in a phenomenon known as the Casimir effect, named after Henrik Casimir, the Dutch physicist who discovered it in 1933. Normally two uncharged parallel metal plates would remain stationary because there is no electromagnetic charge to exert a force to pull them together (or push them apart). But Casimir found that if the plates are close enough, there is still a tiny attractive force between them. Because the parallel plates are so close together, virtual particle pairs can't easily come between the plates, so there are more pairs popping into existence around the exterior of the plates than there are between them. The imbalance creates an inward force from the outside that pushes the plates together slightly. The smaller the separation between the plates, the fewer virtual pairs can get between them, and the greater the force of the inward attraction.

If we could figure out a way to harvest just the antiparticle of a virtual pair, we would have a built-in source of negative energy in the quantum vacuum. Alas, that's a pretty big "if." Even if we could find a way to harvest this negative energy, it isn't remotely sufficient for portal purposes. A wormhole only one meter wide would require negative energy equivalent to the total energy produced by our sun over the course of roughly 10 billion years, yet the Casimir effect is quite small, equal to the weight of 1/30,000 of an ant. Any wormhole resulting from it would have to be much smaller than an atom, making travel through it impractical at best. We would need to find a means of amplifying that energy many times over before it would become strong enough to hold open a macroscopic wormhole.

Frankly, the energy contained in the quantum vacuum isn't nearly as much as Tesla supposed, when one converts it into macroscale units of measurement. Lawrence Krauss estimates that if we could release the energy stored in one cubic meter of the quantum vacuum—an area about the size of a small Dumpster—we would only harvest about one ten-billionth of a joule, not even enough to light a 10-watt bulb. That might make a difference in the subatomic realm, but it hardly seems sufficient to open a dimensional portal, or to perform any other of the impressive magical feats that occur on an almost daily basis in the Buffyverse. No wonder we don't see such exotic phenomena occurring in our own world.

Nonetheless, CalTech physicist Kip Thorne devised a wormhole model based on negative energy. Hypothetically, his model would permit someone to travel not just between different points in space—as with conventional wormholes—but also between different points in time. Thorne proposed creating two identical chambers, each of which contains two parallel metal plates separated by a very small gap. The electrical field created by the plates via the Casimir effect creates a tear in space-time, so that the chambers become the two "mouths" of a connecting wormhole. Then one chamber is placed on a rocket ship and accelerated to near the speed of light. Since time is moving at different rates in each chamber due to relativistic time dilation (remember that a traveling clock ticks more slowly than a stationary one), the two chambers become desynchronized. They are still connected by the wormhole, yet they exist in different times. Time has passed more slowly in the accelerating chamber, so a person in the earthbound

I THINK THAT THIS MIGHT BE TOO SMALL TO USE AS A PORTALLY-TRANSPORTER THINGY.
F
d
F ~ A/d4

chamber could step through the wormhole and be hurtled into the past. A similar effect could be achieved by connecting a wormhole between the earth and something very heavy, like a neutron star. This also sets up a time difference between the two ends, since mass warps space and time. A clock on the surface of a very dense neutron star would run about 30 percent slower than it does on earth.

In the Buffyverse, we wouldn't need rocket ships capable of approaching the speed of light, or massive neutron stars, because time already moves at different rates in other dimensions. Relativistic time dilation is built into the system. *Angel*'s Los Angeles and Pylea each correspond to the mouth of a wormhole, and the two dimensions are moving with respect to each other, as all dimensions do in the Buffyverse. So if the members of Team Angel travel through a portal to Pylea, they are not only traveling to a different point in space, but also to a different point in the past or future, depending on how fast or slow time moves in the respective dimensions. The same goes for Illyria, Jhiera and her compatriots from Oden-Tao, and for any other being that hops between dimensions. They are all unwitting time travelers, whether the difference is 100 years or just a few seconds.

NEGATIVE FEEDBACK

Thorne's wormhole design is quite ingenious; in fact, it formed the basis for Carl Sagan's novel *Contact*. Yet once again, there is a catch. Quantum vacuum fluctuations would almost certainly destroy such a wormhole before it could be used as a portal, thanks to what amounts to a devastating feedback loop, similar to what

Oz's band, Dingoes Ate My Baby, might experience with their guitar amps during a sound check. In such a model, virtual particles pass through the wormhole to the past. But then they must travel forward through space and time, eventually reentering the wormhole and traveling back to the past again in a never-ending cycle. Each time this occurs, the radiation from the particles increases in intensity until it becomes strong enough to destroy any physical object that tries to pass through to the other side. Ultimately it would even destroy the wormhole.

To get an idea of how this works, let's revisit Lindsey's suburban Misner prison, where the boundary "walls" correspond to the mouths of a wormhole. Normally these walls are stationary, but what happens if they are moving with respect to each other, like the different dimensions in the Buffyerse? Imagine that Lindsey's prison is being squeezed to further torment the inhabitants. Anxious to escape his increasingly claustrophobic existence, Lindsey walks repeatedly toward the left wall of his prison, only to find himself reappearing from the right wall. Perhaps he figures that if he repeats the action often enough, a statistical quantum anomaly will occur and he will find himself outside the prison dimension. Instead, he is sowing the seeds of his own destruction. If the right wall is moving toward Lindsey at 2 mph, and he walks through the left wall traveling at 2 mph, when he returns back through the moving right wall, his speed will have increased. He is now traveling at 4 mph, because the two speeds add together. Every time he completes a circuit, his speed increases by another 2 mph. Something similar happens if Lindsey shines a flashlight at the left wall. The beam will gain energy every time it emerges from the right wall, making the Misner prison unstable. The light loops around and around until it becomes so energetic, it creates an enormous gravitational field of its own, which collapses the walls.

Any attempt to use a wormhole as a time machine would give

rise to similar subatomic fluctuations. The wormhole would self-destruct. Because they are also technically wormhole "time machines," the dimensional portals in the Buffyverse should be subject to the same intense radiation feedback loops that plague real-world models of wormholes. A portal should collapse every time someone (or something) tries to pass through it. Even if it didn't, the intense radiation emitted should be sufficient to kill any aspiring dimensional jumpers.

Lindsey's experience of being squeezed inside his Misner prison might seem suspiciously akin to a closed timelike curve, *à la* Buffy's loopy day at the Magic Box or the Russian ballerina condemned by Count Kurskov to dance the same performance for eternity. It should. Some wormhole models—namely, the time-machine variety—can form different kinds of closed timelike curves. In fact, a new type of doughnut-shaped closed timelike curve, proposed in 2005, offers the possibility of a traversable wormhole enabling time travel into the past without the need for such exotic elements as negative matter or energy. Every slice of time after the time machine was created would exist somewhere in the vacuum inside the gravity doughnut's "hole." A traveler in a rocket ship could hypothetically zip around inside the doughnut, receding a bit farther into the past with every completed circuit. Of course, physicists have yet to figure out how to curve space-time sufficiently to form the gravity doughnut in the first place—the energy required would be truly astronomical—and even if they managed to do so, the resulting wormhole would still suffer from the same quantum feedback effects, leading it to self-destruct.

Thus far, the obstacles to building big swirly wormholes have proven largely insurmountable. Yet theorists are a tenacious bunch and are not so easily dissuaded. The search continues for a feasible wormhole model that might one day serve as a portal to other universes, or to other points in time. Various inhabitants of the Buffyverse have clearly outpaced real-world physicists in solving the dilemma, even if they don't fully understand the underlying physical mechanisms behind the magic. We must look to Fred, the intrepid physics student, and her post-Pylean foray into string theory, for further insights into this exotic feature of the Buffyverse.

9

SHATTERED SYMMETRIES

Stringing Together a Theory of Everything

"There is no such thing as perfection. The world itself is imperfect. That's what makes it so beautiful."
—Colonel Roy Mustang, FULLMETAL ALCHEMIST

Itching to get back into physics after her five-year exile in Pylea, Fred publishes a scholarly paper on "Superstring Theory and *p*-Dimensional Subspace," which causes quite a stir in the local scientific community. As a result, she is invited to give a presentation on her findings at a special colloquium held at the fictional California Physics Institute. She purportedly shares the stage with none other than noted real-life string theorists Edward Witten and Brian Greene. While other budding young physicists struggle to build their reputations in the minor leagues of science, Fred appears to have leapfrogged straight into the majors.

"Supersymmetry" (A-4) is perhaps the most overtly physics-intensive episode in the Buffyverse oeuvre—the other candidate being Gene's time-stopping activities in "Happy Anniversary" (A-2). Let's overlook the unlikelihood that an unknown physics student would ever share a podium with scientists of Greene's and Witten's stature. Witten is one of the top string theorists in the world, and Greene is the field's greatest ambassador; his bestselling book *The Elegant Universe* propelled string theory into the popular imagination. Fred's dialogue in other episodes is peppered with vague scientific references that, when parsed, often don't make a lot of sense. But in this case, she's not just spouting random technobabble when she delivers her lecture. D-branes, *p*-dimensions, T-dualities, Dirichlet boundary conditions, and compactification are all bona fide terms in string theory, and Fred uses them correctly—not that anyone who wasn't a string theorist could tell the difference offhand. And while the journal in which Fred's paper appears, *Modern Physics Review*, is fictional, its name bears an uncanny resemblance to *Reviews of Modern Physics,* one of the most prestigious real-world physics journals. So give the writers due credit for doing their homework.

To a layman's ears, the subject of Fred's paper sounds impossibly abstract and esoteric, and it should: It addresses one of the most ambitious long-term objectives in physics. Real-world physicists are indeed searching for a theory of everything (TOE), "something that combines quantum mechanics and general relativity, accounting for both the behavior of the smallest subatomic particles and the largest forces of nature," as Fred puts it. A bona fide TOE would describe how the universe works at all size scales, instead of having two competing sets of rules for the separate realms of small and large. String theory is the top contender to date for that honor.

HIGHER LEARNING

Fred arrives at the institute to deliver her lecture and soon encounters her old academic advisor, Oliver Seidel, who reminisces about how quickly Fred's aptitude for physics became apparent in his introductory physics course: "By the end of the semester, she was taking on WIMPs!" A confused Gunn naturally associates the term with weak-kneed milquetoasts, and boasts of his girlfriend's budding pugilistic abilities. Fred gently explains that the acronym stands for weakly interacting massive particles, one of the candidates for the strange dark matter that physicists believe makes up the bulk of matter in the universe.

WIMPs may be mentioned primarily as comic relief, but they are a very real, very active field of physics, and any first-year physics student capable of studying them would indeed be quite exceptional. Her scientific credentials thus established, Fred makes an appointment with Seidel the next day to discuss some of the implications of her new theory, specifically how it fits with "Kaluza-Klein models." This is a direct reference to an early-twentieth-century attempt to unify gravity and electromagnetism.

Einstein unified three-dimensional space with the fourth dimension of time, and merged gravity and acceleration, attributing the force of gravity to the warping of the fabric of space-time. He dreamed of one day unifying the strong and weak nuclear forces as well. In 1919, a Polish mathematician named Theodr Kaluza proposed that another force, electromagnetism, might be due to a similar warping of an unseen fifth spatial dimension. By reworking Einstein's equations in five dimensions (four spatial, one temporal) instead of four, Kaluza believed that he could merge the two forces of gravity and electromagnetism. Light could therefore be envisioned as a disturbance caused by the rippling of the higher dimension just beyond human perception, much as fish in a pond can only see the shadows of the ripples across the water's surface caused by raindrops.

This is a difficult concept for most of us to visualize, since we live in a three-dimensional world. The prospect of a higher spatial dimension is beyond our limited perspective. Most of the inhabitants of the Buffyverse share a similar limited world-view. However, Cordelia eventually makes so much progress in her personal development while working for Angel Investigations that the Powers That Be elevate her to the status of Higher Being ("Tomorrow," A-3). In her new incarnation, Cordelia ascends to a higher dimension, where time's arrow no longer exists, making her immortal. Stuck in their lower dimensional reality, the members of Team Angel are mystified by Cordelia's disappearance.

Let's imagine for a moment that Cordelia is a tiny person living in TV Land instead of an electronic representation of an actress playing a role. Her perspective would necessarily be limited to the TV world of the Buffyverse—whatever she can touch, see, smell, taste, and hear around her in any given episode. She would have no concept of anything outside that boxed-in reality. But if she were suddenly lifted from the television screen, and landed, life-size, in our living room, she would see the characters and events in the Buffyverse unfolding from an entirely new perspective. In fact, she'd be largely omniscient, with the all-seeing eyes of the divine. The same concept applies to the dimension of Higher Beings.

Add in the immortality, and it sounds like a decent gig, but Cordelia is quickly disillusioned and utterly bored with her new

width
height
height
depth
width

role. Since there is no time in this all-seeing realm, nothing ever changes. Plus, she is unable to interact with her former world, or interfere in any way to alter the course of events therein, much like frustrated television viewers are helpless to change how events unfold on their favorite programs. All they can do is change the channel. Ghost-Spike faces a similar dilemma when he finds himself flitting back and forth between the "reality" of Los Angeles and a separate spectral dimension in "Hellbound" (A-5). We've seen how if that dimension consisted of four spatial dimensions instead of three, Ghost-Spike would be able to see and hear Fred and the rest of Team Angel, but they wouldn't be able to see and hear him. And this does indeed seem to be the case. Like Cordelia, Ghost-Spike is aware of what is happening in that lower dimension, but he can't interact with it in any meaningful way.

There is other internal evidence that this spectral realm has more than three spatial dimensions: Ghost-Spike disappears from view and then reappears at random moments as he pops in and out of the higher dimension. At least that's how it seems to Team Angel. In the real world, we lack such definitive evidence, so Kaluza's theory raised an obvious question when it was first proposed: If there is a fifth dimension, why don't we experience it in the same way that we experience our four-dimensional spacetime? Case in point: If one of Fred's scientific experiments causes a small explosion, the resulting smoke will gradually disperse throughout the room until its molecules reach a state of equilibrium with the rest of the molecules in the air. The smoke molecules never magically disappear the way Ghost-Spike does when popping off into the spectral world, but they should if the smoke could seep into a higher dimension.

That's where the "Klein" in Kaluza-Klein theory comes in. Oskar Klein was a Swedish mathematician who found a solution to the conundrum. He argued that this hypothetical fifth dimension could simply be so tiny that not even the atoms that comprise the smoke from Fred's explosive experiment could pass into it. According to Klein's calculations, it would have to be curled up ("compactified") into a tiny ball much, much smaller than an atom. This size scale is known as the Planck length.

Kaluza-Klein models experienced a resurrection of sorts in the 1970s, when string theorists adapted this extradimensional approach to unify not just gravity and electromagnetism, but the strong and weak nuclear forces as well. It's not an easy task, since the four fundamental forces differ greatly in terms of strength. Add in the mishmash of 57 or so subatomic particles, and it calls to mind the bickering horde of disparate demon species that make up the bulk of Wolfram & Hart's clientele. The firm's lawyers are masters at smoothing over disputes and finding ways to get their clients to work together. Similarly, some kind of mediating model is needed to bring together all the disparate elements of the cosmos.

Over the course of the twentieth century, physicists painstakingly cobbled together a reasonably efficient "standard model" of physics. The model they came up with almost works without resorting to extra dimensions. It merges electromagnetism with the strong and weak nuclear forces (at almost impossibly high temperatures) despite the differences in their respective strengths, and provides a neat theoretical framework for the big noisy "family" of subatomic particles. But there is a gaping hole. The standard model doesn't include the gravitational force. That's like building a model incorporating all the characters and various elements of the Buffyverse, but inexplicably leaving out vampires. Vampires, like gravity, appear to be strong from the perspective of the humans who populate the Buffyverse, but they are surprisingly weak compared to some of the other demons (their sensitivity to sunlight, for example, puts them at a disadvantage), just like gravity is weak compared to the other fundamental forces. Yet vampires are an integral aspect of the Buffyverse. One ignores them at one's peril. The same is true of gravity. It's the proverbial thorn in the standard model's side, the pink polka-dotted elephant—or fanged, yellow-eyed bloodsucker—in the middle of the room that nobody likes to mention. It's the piece that doesn't quite fit into the rest of the jigsaw puzzle.

In contrast, string theory incorporates gravity quite elegantly. Einstein's equations emerge naturally from the broader mathematical framework. But it achieves its grand unification at a price: It requires extra dimensions. Much like Cordelia ascending to a higher plane of existence, string theory extends Kaluza and Klein's basic approach to dizzying new heights, calling for a whopping nine dimensions of space and one dimension of time. That's how many it takes to merge the four fundamental forces into a single TOE, because our own four-dimensional space-time simply doesn't have enough room to accommodate them all. And just as with Kaluza-Klein models, those extra dimensions may be compactified down to the subatomic scale, and can't be seen.

PULLING STRINGS

The crux of Fred's paper concerns the notion of string compactification, which is tied to the crumpled-up extra dimensions of space we've just discussed. To quote Fred, "If space-time can undergo massive rearrangement of its structure—and I believe it can—tearing and reconnecting according to a predetermined disposition, then T-duality would allow for the compactification of extra spatial dimensions." Stripped of the jargon and translated into everyday English, it appears that Fred is looking to this aspect of string theory to explain the mechanism behind dimensional portals in the Buffyverse. Indeed, she credits her experiences in Pylea with inspiring her new insights into string theory: "If I hadn't been sucked into that portal, I never would have figured out my string compactification theory." Most of us would respond to that statement much like Gunn does in the episode: "Exactly, 'cause, you know, strings . . . need . . . to compactify."

What exactly are these mysterious strings, and how could they possibly have anything to do with portals in the Buffyverse? There are no simple answers to these questions, but we can start by recapping what we've learned thus far about black holes. By their very nature, black holes are a paradox. The center of a black hole is both massively heavy and tiny at the same time, and a black hole may even have a wormhole at the center connecting it to another point in space-time—the conceptual foundation for dimensional portals in the Buffyverse. Both quantum mechanics and relativity therefore apply, yet scientists can't apply two such vastly different theories simultaneously. A master set of equations (a TOE) that could describe the universe at all size scales would enable physicists to do just that, and hopefully resolve the remaining mysteries surrounding black holes, including the question of what really lies at the heart of one.

One of the critical problems in merging the two theories is that relativity requires a space-time fabric that is smooth. But at the quantum level, space-time is bumpy, with virtual particles popping in and out of existence like tiny bubbles frothing in a subatomic foam. Quantum fluctuations are so small that they don't matter at the macroscale. They become much more significant at the subatomic level, which is why Newton's laws of motion and Einstein's relativity—both of which describe gravity—break down on that tiny size scale. String theory attempts to address this contradiction by reenvisioning the nature of subatomic particles, which are traditionally viewed as tiny, fixed points.

Thanks to the uncertainty principle, if we know a particle's exact location, we can't know anything about its velocity (or momentum). That's why we can't use Newton's laws to determine the trajectory of a subatomic stake. Classical physics demands that we know the values for both properties. But what if these pointlike particles are, instead, tiny vibrating strings: one-dimensional closed-loop filaments akin to infinitely thin rubber bands? Perhaps strings only appear to be fixed and pointlike because they are so small. If they are vibrating, our knowledge of their position is not so exact. This fuzziness calms the so-called quantum jitters.

According to string theorists, there are the three full-size spatial dimensions that we experience every day, one dimension of time, and six extra dimensions crumpled up at the Planck scale like itty-bitty wads of paper. As tiny as these dimensions are, strings are even smaller. Fred, Gunn, and the rest of Team Angel (and us) are limited to four-dimensional space-time. A string enjoys the same kind of unfettered freedom as a Higher Being in the Buffyverse, able to move through all ten dimensions of space-time. But unlike the frustrated Cordelia, who is unable to interfere with the mortal realm during her yawn-inducing stint as a Higher Being, a string's multidimensional existence has a very real, observable effect on our physical reality. The little string wriggles as it goes on its merry way, and the geometric shape of those extra dimensions helps determine the resonant patterns of vibration. Those vibrating patterns in turn determine the kind of elementary particles that are formed, and they generate the physical forces we observe around us—in much the same way that vibrating fields of electricity and magnetism give rise to the entire spectrum of light or vibrating strings can produce different musical notes on a violin. All matter, and all forces, are composed of these vibrations.

Impressive, you might be thinking, but what about those portals? This brings us back to wormholes. In order to create a wormhole, the fabric of space-time must rip, yet there is still some debate among physicists as to whether this is possible. And once again, the two prevailing "rule books" disagree. General relativity says that space-time can bend and curve, but it doesn't allow for punctures or tears; space-time must be smooth. Einstein speculated on the possibility of preexisting wormholes at the heart of black holes—akin to the preexisting "hot spots" that serve as natural gateways to other worlds in the Buffyverse—but nixed the idea that we could one day create a new one. However, quantum fluctuations can give rise to hypothetical wormholes at the subatomic scale: tiny tears in space-time that exist for fractions of a second before closing up again.

Here's what keeps a theoretical physicist awake at night: If space-time can rip, what's to prevent that hole from getting bigger, until the entire fabric is split in two? This would be a cosmic catastrophe, as we shall see in the next chapter. Fortunately, the universe appears to have a built-in "cosmic censorship" principle at work to guard against these events, known as "naked singularities." Black holes provide a safeguard in general relativity. Since nothing can escape their immense gravitational fields, the hypothetical punctures at the singularity are cut off from the rest of space-time, so the rip is unable to spread uncontrollably. And at the quantum level, the tears exist for mere fractions of a second, so there just isn't enough time for rips to reach catastrophic proportions.

String theory offers its own unique solution to the dilemma. It allows space-time to puncture or tear as it compactifies, but it can also repair itself and "reinflate," often into a new configuration. Crazy though it may sound, Fred's theory might just be on the right track. Among their many other useful properties, strings can compactify and reinflate along with a shrinking region of space. The string wraps around the spatial area, just like one could wrap a rubber band around an object. Because it's so stretchy, the string's ability to expand and contract in size provides a protective "shrink-wrapping" around any torn patch of space-time. This keeps the tear from ripping out of control, until that bit of space can repair and reinflate.

That's the working hypothesis, anyway, and it is certainly borne out mathematically in the few specific configurations for which the numbers have been crunched. These have been for the curled-up, six dimensions of space, but there's no reason the same rules can't apply to our own four-dimensional space-time—although wormholes (traversable or not) remain a distant hypothetical possibility, even in string theory. The lack of direct experimental evidence for wormholes turns out not to be a problem for Fred. Shortly after she begins her lecture, a portal opens up right behind her on the podium—the kind of dramatic, undeniable, eyewitness proof every physicist dreams of one day finding to support his or her pet theory.

IT TAKES BRANES

Not only does string theory offer a convincing mechanism for portals in the Buffyverse, it can also account for the origin of its myriad dimensional worlds. When Fred meets with Seidel the day after her horrific close call, she explains about her exile to Pylea, prompting bemused skepticism in the physics professor. It's a perfectly understandable reaction from a scientist who has witnessed

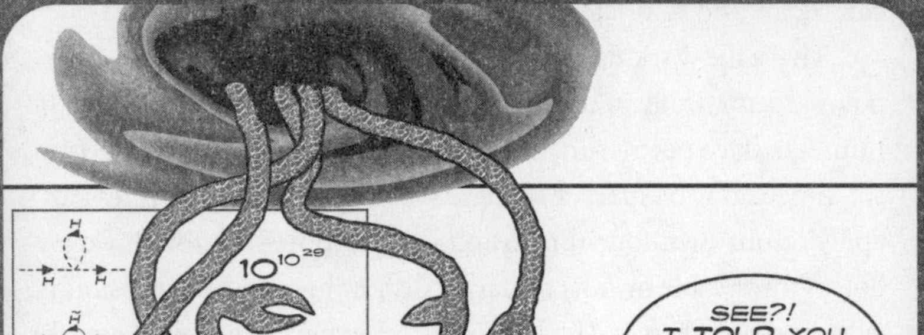
SEE?! I TOLD YOU THAT SPACE-TIME COULD RIP!!
10^10^29
Mp = 2p-1
Mp = 2p-1Mp

a truly otherworldly occurrence but is still cautious about drawing wildly speculative conclusions. He's more inclined to attribute the portal's unexpected appearance to subconscious suggestion or a form of mass hysteria. As for Pylea, Seidel admits to Fred that, as a theoretical physicist, he is open to the possibility that other worlds exist, "but you're naming them."

Of course, when Fred speaks of other dimensions in this context, she does not mean the crumpled-up, tiny extra dimensions of string theory. Because they are so tiny, those dimensions are not the sort that might house ghosts, demon races, Higher Beings, or shy physics students who've been sucked through a portal by mistake. So those dimensions can't intermingle with the four dimensions of space-time in which we exist, and we can't, in turn, enter them. Fred is using the term to describe the equivalent of parallel universes. The Buffyverse, as we have seen, is a multiverse. All those parallel worlds, like Pylea, may have started out as baby universes, budding off from tiny wormholes that popped up in the quantum foam before inflating to their full size. It just so happens that the latest incarnation of string theory offers an alternative explanation for the origin of parallel universes.

Strings are really stretchy, so hypothetically a tiny string can stretch into an object resembling a membrane. String theorists have dubbed these structures "branes." Unlike strings, branes might be quite large, but still invisible to the standard ways scientists probe the universe. A brane can exist in any dimension less than ten. Point particles are "zero branes." A string is one-dimensional, like a line, so it is called a "one-brane." A membrane is two-dimensional, defined only by length and width; this is a "two-brane." If a brane has length, width, and breadth (three spatial dimensions), it is a "three-brane." And so forth. The so-called *p*-branes are simply scientific shorthand for denoting various possible dimensions for branes without specifically designating one; *p* can equal any whole number less than 10. The same concept applies when Fred talks about *p*-dimensional subspace.

So not all extra dimensions must be wrapped into a tight little Planckian ball. Some can be huge, as large as a universe. In fact, it's possible to view our entire universe as a gigantic three-brane floating in a much larger dimensional brane-world, much like the world of the Buffyverse exists in TV Land, which is in turn part of our much larger living-room reality. Brian Greene likens it to living inside a loaf of bread, where our universe is just one slice in a much larger loaf. There could be parallel universes floating right next to ours as little as one millimeter away, and we are simply oblivious to their existence. We're back in multiverse territory, with other worlds—like Pylea, or Quor-Toth—that may, or may not, resemble ours in terms of physical laws.

Instead of springing up out of the quantum foam and budding off the parent universe, perhaps parallel universes arise from collisions between large brane-worlds. This is admittedly highly speculative, even among string theorists, but it's an intriguing notion. What happens when two giant branes floating in hyperspace collide? The working hypothesis is that the kinetic energy from any such collision would be so immense, it would convert into matter and energy, much like the spark from a flint striking stone can start a fire. The intense temperatures produced by the collision would then cool rapidly as the two membranes move apart.

So our universe might be the result of a "big splat" instead of a big bang. And it could happen again in the distant future, because the expansion of our universe is accelerating. Physicists aren't sure why this is the case, but many hypothesize that it's due to the existence of "dark energy," a type of antigravity (similar to the negative energy discussed in the previous chapter) that is pushing the galaxies farther and farther apart. But there is another option. If the colliding-branes scenario were to prove correct, there is no need for dark energy. The accelerated expansion rate could be due to the gravitational attraction that still exists between our brane and a second brane with which we once collided. The two branes could once again be moving toward each other at an ever-faster rate, which means another collision could happen billions of years from now, vaporizing our world in a cosmic fireball even as it creates another baby universe in our stead. In fact, such collisions could have happened countless times, giving rise to any number of nascent universes.

We might not be able to see these parallel universes, and we certainly can't travel to and from those other worlds via mystical portals as in the Buffyverse. But perhaps one day we will be able to sense them through gravity. Among its many other insights, string theory offers a potential explanation for why the gravitational force is so much weaker than the other forces. Originally string theorists conceived of strings that were closed loops, like rubber bands, able to travel freely among the higher dimensions. But now they believe that there could also be open-ended strings, where each endpoint is tied down to a brane. Such strings have set boundaries. They are limited to just those dimensions of the particular branes to which they are attached—the "D-branes" Fred mentions in her lecture.* Much like Team Angel is confined to the three-dimensional world of Los Angeles, matter and light would be made of open-ended strings, and are thus confined to our three-dimensional D-brane. The same goes for the "messenger" particles (bosons) that carry the weak and strong nuclear forces. These are the ways in which physicists normally explore the universe.

*The "D" stands for "Dirichlet" (pronounced "deh-RISH-lay"). Johann Peter Gustav LeJeune Dirichlet was a nineteenth-century mathematician known for applying certain boundary conditions to differential equations. Branes that serve as "sticking points" for open strings set similar boundary conditions for those strings. That's why they're known as D-branes.

In contrast, gravity might be more like a Higher Being in the Buffyverse. A closed-loop string that has been dubbed a graviton—the hypothetical messenger particle for gravity—has no loose ends that are tied onto a brane, and hence no boundaries, so gravitons can escape into higher dimensions. It could be that gravity isn't such a weakling after all. We just can't feel its full impact, because it dissipates into extra dimensions. The atoms and particles of our universe stay on our brane, in much the same way that Spike's billiard balls stay on the surface of the pool table in the Bronze. But whenever he makes a shot, balls collide and sound waves—mechanical vibrations in the air—travel off the table and into the surrounding atmosphere. Other people can hear the collision even if they aren't in the same room. Gravity, too, might be able to escape the confining parameters of its source and seep into other dimensional "rooms," thus diluting its strength.

Physicists might one day be able to detect interference patterns from these hypothetical branes, encoded in gravitational waves. These are ripples in the fabric of space-time caused by violent events in the universe: colliding black holes, shock waves from exploding supernovae, perhaps even the remnant of the shock wave produced by the big bang itself. Of course, first we must be able to sense gravitational waves, which are very faint. Any information encoded in those ripples would be even fainter. The good news is that scientists may be on the verge of observing gravitational waves directly for the first time with the Laser Interferometer Gravitational-Wave Observatory (LIGO), which began operation in 2005, as well as the space-based Laser Interferometer Space Antenna (LISA), slated for launch around 2015. Should they succeed in observation, even the skeptical Seidel might be forced to accept the existence of extradimensional brane-worlds as a real scientific possibility.

CALCULATED COUPLINGS

It's not branes, but bodies that collide in "Supersymmetry" (A-4). The episode opens with Fred and Gunn making love to celebrate the publication of her paper. And it's not the only romantic coupling that takes place. Wesley is mired in a passionate entanglement with Lilah Morgan, a high-powered lawyer with Wolfram & Hart. And a budding romance between Angel and Cordelia has been complicated by the growing attraction between Cordelia and Angel's eighteen-year-old son, Connor, setting up an awkward (and slightly incestuous) love triangle. Who knew string theory could be so steamy? It just so happens that string theory contains its own elaborate network of couplings, in the form of "stringy dualities."

Dualities in physics refers to theoretical models that appear to be different but can be shown to describe exactly the same thing. It's a bit like how ice, water, and vapor are three different phases of the same chemical substance, except that a duality looks at the same phenomenon in two different ways that are inversely related. Remember the kindly Sunnydale doctor, Ben, and his hell god alter ego, Glory? We learned in chapter 6 that subatomic particles like photons and electrons also exhibit wavelike behavior, but uncertainty dictates that we can't see both of these aspects at the same time. The more accurately we observe its particle nature, the less we can see the wavelike nature of the object, and vice versa. This is a duality. The particle/wave identities are inversely related, different "faces" of the same underlying reality, just like Ben and Glory are inverse (male and female) personas inhabiting the same body.

String theory has its own set of dualities. Fred rightly says that there are several competing dimensional theories—five, in fact—and as recently as 1995, no one knew which version was correct. Then Edward Witten stepped in and demonstrated that the five different string theories weren't so contradictory after all. He united all five under a single theoretical umbrella that he dubbed M theory,* adding a tenth spatial dimension to bring the total number of required dimensions to eleven.

M theory indicates that mathematically, the five versions of string theory are merely five different ways of looking at the same thing. Just like the couplings that take place between the various characters in the Buffyverse, each iteration of string theory is connected in some way to another through various dualities in an intricate web of interconnections that ultimately links all five to one another and to M theory. But whereas the complex social network of relationships in the Buffyverse usually leads to conflict and broken hearts, stringy dualities can help physicists simplify difficult calculations through a kind of "bait-and-switch" approach. Fred is referring to these intrinsic dualities when she asks her audience to "consider the nonperturbative properties of string theory" during her lecture.

To understand what this is, we must first understand what it isn't. Perturbation theory is a calculating method that makes approximations to get a rough answer and then refines it bit by bit, according to how given physical systems are known to behave. Let's say that Xander, in his role as construction manager, makes an initial estimate for a prospective client to rebuild Sunnydale High School. Xander can then gradually make refinements to that estimate by filling in the missing details as they emerge to calculate the final bill: the actual materials and labor costs, for example.

*Nobody knows for sure what the "M" stands for. Possibilities include "mother," "membrane," "matrix," "mystery," even an inverted "W" (for Witten). Witten himself isn't telling.

These details are "perturbations" to the original estimate. Physicists employ a similar approach when determining the trajectory of a satellite. They use Newton's laws for the initial calculation and then make small refinements by calculating the effects of other, minor factors that might influence the trajectory: pressure from the solar wind, for instance, or the effects of heating on one side of the satellite.

Ideally, there should be only small discrepancies between Xander's original estimate and the final bill, just as there should be only small discrepancies between the predicted properties of a physical system and those that actually emerge. But as anyone who has ever hired a contractor knows, the final bill is sometimes way off target. Complications may ensue. There could be massive cost overruns from a labor strike, rising materials costs, or unexpected damage from demon attacks or marauding students driven mad by the percolating Hellmouth energy below the school. In those cases, the refinements result in large changes to the ballpark estimate: "contract-y goodness" for Xander, who stands to make a tidy profit from the overruns, but a financial disaster for the client, who probably didn't budget for those eventualities.

Something similar happens when physicists try to calculate the highly turbulent air flow patterns of a tornado, for instance, or the properties of a rapidly rotating wormhole. Their perturbative approximations break down because they are dealing with sudden, large changes to the original value, instead of small, predictable increments. If Fred were trying to calculate the properties of a swirling dimensional portal in the Buffyverse—which would be similar to the physics of a rotating wormhole—she wouldn't be able to use a perturbative approach, because the numbers involved would be too large. Stringy dualities could help simplify her calculations.

Among the jargon Fred employs in her lecture is something

called T-duality. Because it's a duality, we know it describes an inverse mirror relationship between two of the five permutations of string theories. It also pertains to transformations of space-time, a critical requirement for the creation of a wormhole or portal. Here's how it works: if the radius (r) of a circular area of space has a large value ($r = 1{,}000$) in one of the five versions of string theory, it is difficult for Fred to calculate. However, that value will be inverted, and therefore small ($r = 1/1000$) in one of the other versions. This is a small perturbation, and easier for Fred to calculate. Because both theories describe the same underlying physics, Fred can use the dual theory where the value is small to calculate the quantity, then plug it back into the original theory to complete the calculation. No doubt that's how she arrives at her revolutionary breakthrough in string theory (as it exists in the Buffyverse): Space-time can shrink, tear, repair, and reinflate, thereby rearranging itself into exotic configurations like portals to other dimensions.

Dualities are a type of symmetry, and it's no accident that this episode is entitled "Supersymmetry." The concept is a natural extension of the many different kinds of symmetry we see every day around us in the physical world. For instance, rotate a snowflake by 60 degrees, and you'll find that it looks exactly the same. This is spatial symmetry. A second type of symmetry occurs when one shuffles a series of similar objects, as in a shell game, where the player must guess under which of three shells a marble might be found after the three are randomly shuffled in quick succession. Regardless of where the marble turns out to be, mathematically there are six different ways in which three identical objects can be interchanged. The equations that appear in the background in the episode don't come from string theory, but from quantum chromodynamics (QCD), a theory that describes the strong nuclear force and the way various quarks interact with one another. There are quarks of three different "colors" that can be randomly interchanged, just like the shells, so those quarks share a similar internal symmetry.

Supersymmetry extends this interchangeable shuffling to incorporate all known subatomic particles. Not all potential couplings are feasible in the current standard model. Fermions (the particles that make up matter) and bosons (messenger particles that carry fundamental forces) can't mix at all because they have such vastly different properties. It's a bit like how certain demons in the Buffyverse have "anatomical peculiarities" that prevent them from mating with any creature outside their species—at least not without causing severe internal damage. Supersymmetry allows us to engage in "cross-species" mating, interchanging a fermion with a boson. In order to accomplish this, however, it requires some sort of "adapter"—in this case, the existence of hypothetical super partners, called sparticles. Each fermion is paired with a super-boson partner, and each boson has a super-fermion partner. Now they can be mixed via their super partners, but the price is a doubling of the number of subatomic particles. Never one to shy away from a little added complexity, string theory incorporates supersymmetry.

If it turns out to be true, supersymmetry would provide physicists with a powerful calculating tool for understanding the most elusive mysteries of our universe, since everything would be connected to everything else through various kinds of dualities. Fred gained revolutionary new insights into Buffyverse physics using just one kind of duality. Imagine the major theoretical breakthroughs she could make were she able to simplify all her most difficult calculations in this way. The same goes for real-world physicists. Alas, supersymmetry is broken. Our universe today is an imperfect shadow of its former majestic self.

THE MIRROR CRACK'D

As Fred waits in Seidel's office, she notices that a textbook, *Plasma and Fluid Turbulence,* is inexplicably sandwiched between books on neutrino physics on the shelf. When she opens it, she finds ancient pages from a mystical text with instructions on how to open a dimensional portal. Horrified, Fred realizes that her beloved mentor has been lying about his professed ignorance regarding other dimensions. He is the one responsible not only for opening the portal during her lecture, but also for sending her to Pylea six years earlier.

It seems he's even figured out how to open portals wirelessly. As a vengeance-minded Fred confers with Wesley over a suitable punishment for Seidel, she receives a call on her cell phone. When she flips it open, strange symbols—no doubt the "consonant representations of mathematical formulae" with which she has become so familiar—appear on the display screen, and a portal opens up right there in the room. That makes Seidel pretty cutting-edge in terms of real-world technology. Wireless devices have become mainstays of our daily lives, from cell phones, BlackBerries, and laptops to embedded wireless sensors in buildings and bridges to monitor structural integrity. Perhaps he should have been an experimentalist instead of a theorist.

Fred survived her five-year ordeal in Pylea, then built a well-ordered life in Los Angeles as part of Team Angel. But by the episode's end, her carefully structured world has come apart at the seams. The symmetry has been broken, and she must once again piece the broken shards of her life back together. It's a useful metaphor for the supersymmetry that string theorists believe characterized the early universe. Before the big bang, the cosmos was a perfectly symmetrical ten-dimensional universe with all four fundamental forces unified at unimaginably high temperatures—a state of extremely low entropy. But this universe was highly unstable and cracked in two, sending an immense shock wave reverberating through the cosmos. The result was two separate space-times: the unfurled four-dimensional one that we inhabit, and a six-dimensional one that contracted as violently as ours expanded, shrinking into a tiny Planckian ball. As our universe expanded and cooled, the four forces split off one by one, beginning with gravity. Everything we see around us today is a mere shard of the original shattered ten-dimensional universe.

Physicists aren't sure why it happened, but they suspect that it might be due to the incredible tension and high energy required to maintain such a highly ordered supersymmetric state, which could render it inherently unstable. Imagine that Fred and Gunn are making their shared bed on laundry day, but the bedsheet has shrunk slightly in the wash. They manage to get it to fit around all four corners of the bed, but the sheet is stretched so tightly that it just won't stay in place. There is too much strain on the fabric, so one corner inevitably pops loose, causing the bedsheet to curl up in that spot. Fred and Gunn can force that corner back into place, but again, the strain will prove to be too much and another corner will pop. Like the bedsheet, the original ten-dimensional fabric of space-time was stretched tight in a supersymmetric state. But the tension became too great, and space-time cracked in two. One part curled up into a tight little ball, while the aftershock from the cataclysmic cosmic cracking caused the other part to expand outward rapidly. This became our visible universe.

For a ten-dimensional universe, there are apparently millions of ways for supersymmetry to break, although there does seem to be something special about three spatial dimensions that causes that configuration to be favored. We've seen that the way the

$(\frac{1}{2},0)\otimes V\oplus(0,\frac{1}{2})\otimes V^*$

remaining dimensions curl up (their shape) gives rise to all the fundamental constants, the respective strengths of the various forces, and ultimately the structure of the universe. If we apply symmetry breaking to the multiverse—or the many different parallel dimensions that exist in the Buffyverse—it gives us a potential explanation for why the laws of physics may (or may not) be different in such universes. Each time a baby universe forms, whether it arises from the quantum foam or from the energy of colliding branes, the symmetry will break randomly in a different dimensional configuration, which in turn determines what kind of universe is born.

Scientists still don't understand the origin of symmetry breaking, but on a less-cosmic scale, some version of it appears to be a crucial component in many basic physical processes, including simple phase transitions, such as the critical temperature/pressure point where water turns into ice. In fact, some kind of symmetry breaking is woven into every aspect of our existence, right down to the mutual (one assumes) orgasms Fred and Gunn experience while making love. The sexual energy between them builds and builds, until finally the tension becomes so great that something quite literally has to give.

Paradoxically, it is shattered symmetries that make life possible, and not just in the context of human sexuality and reproduction. Remember that tiny amount of extra matter that led to its victory over antimatter, and gave rise to everything in our universe? That's just one example of symmetry breaking that proved critical to our existence. It's called CP symmetry, where C = charge and P = parity. All this means is that matter and antimatter have opposite charges, and there should be equal amounts of each—except there wasn't. That tiny asymmetry in favor of matter literally determined our fate.

The same is true of time. We've seen that all the major physics theories exhibit time-reversal symmetry: Time can flow in both directions. At the subatomic level, this does indeed appear to be the case. If you could videotape a particle and antiparticle popping out of the quantum vacuum and annihilating into radiation, and then ran the tape backward, it would be impossible to tell the difference between the two directions. It's a quantum palindrome. Yet in our macroscale day-to-day world, the symmetry of time has been broken. Entropy holds sway. For us, time is a one-way street.

That's not necessarily a bad thing. Nothing could ever change if there were no distinction between the forward and backward motion of time. Not only would we be bored out of our minds by the relentless sameness of our surroundings, like Cordelia, the reluctant Higher Being, there wouldn't be much of a universe at all, because nothing could ever permanently form. Think about it: if an object forms, it creates a distinction between past (when there was no object) and present (in which the object now exists), which means we can tell the difference between time moving forward or backward. Human memory serves the same purpose. We can remember the past, and this establishes a time line moving in one direction: forward. The downside, as Wesley warns Fred about the price of vengeance, is that "once you've acted, you can't go back. You'll have to live with your actions forever."

Scientists aren't sure about the origin of time's arrow, but they found a potential clue with the discovery of an extremely rare elementary particle called the kaon. As elementary particles go, the kaon has a much longer life span—still mere fractions of a second, but those fractions are significantly greater than for the rest of the subatomic family. And unlike its subatomic siblings, the kaon doesn't exhibit time symmetry (nor does another particle called the B meson). As with matter and antimatter, there is a very slight asymmetry, favoring the forward rather than backward motion of time. The kaon and the B meson could hold the secret to time's arrow, which is why they are the focus of such intensive research.

Oddly, the broken symmetry of time's arrow appears to be linked to heart health. In 2005, a team of scientists from Harvard Medical School and the University of Lisbon discovered that too-perfect symmetry in human heartbeats—a cardiogram that looks the same whether run forward or backward in time—is one sign of an unhealthy heart. Young, healthy study subjects showed the most asymmetrical heartbeats.

Unfortunately, symmetry breaking leads not to heart health, but to heartbreak for poor Fred. The pernicious physics professor is sucked into one of his own portals, but not before Gunn ruthlessly snaps his neck. The exacting of vengeance rips the couple apart. Their personal symmetry is shattered just like our nascent cosmos. And as Wesley forewarned, they can't go back to their former idyllic state.

TRIAL BY FIRE

Fred and Gunn's relationship is put to the test and, ultimately, it fails. String theory is undergoing its own intellectual trial by fire, facing mounting criticism from others in the physics community who believe that the theory is inherently untestable. The crux of the criticism is valid. Science currently lacks the means to directly observe strings, so their existence, and that of any extra dimensions, can only be inferred from the underlying mathematics. As with the Many Worlds hypothesis, if a theory can't be tested, it is not so much a science as a philosophy—at least in the minds of most physicists—no matter how impressively complex and elegant the math.

That's why the focus of string theory is shifting more toward developing testable predictions—those that can be verified by practical experiments. The problem is that the energies required to test any of string theory's predictions are literally astronomical. A particle accelerator roughly the size of our own solar system might do the trick, but we can't build something of that magnitude. String theorists have countered that certain predictions could conceivably be tested within the next decade if plans for the next generation of particle accelerators continue on schedule.

The Large Hadron Collider (LHC) now being built in Switzerland should be operating by 2008. Remember Spike's colliding billiard balls? When the balls collide, some of their kinetic energy will be transformed into heat or noise upon impact, but energy conservation dictates that if we add up all the different kinds of energy, there should be the same amount after the collision as there was before. If experiments at the LHC show that there is less energy after a high-energy collision between subatomic particles than at the start, it would constitute an apparent violation of the first law of thermodynamics. But just as mechanical sound waves from the collision of Spike's billiard balls can seep into other rooms, any shortfall in energy following a particle collision could be attributed to some of that energy seeping into extra dimensions—strong circumstantial evidence of their existence. Likewise, if evidence is observed for any of the supersymmetric sparticles, that, too, would bolster string theory's reputation. Discovery of the hypothetical graviton is also a long shot, but this would be another evidentiary feather in string theory's cap.

Not all the criticism of string theory rests on testable predictions. Elegant though it may be, string theory makes one very big assumption: the preexistence of the fabric of space-time. It's a bit like a master painter who produces great works, yet never stops to wonder where the canvas comes from. In truth, referring to space-time as a "fabric" is somewhat misleading. It isn't canvas, cotton, silk, or some cheap polyester blend (a prospect the nattily dressed Lorne would find horrifying). It isn't any kind of material at all, although physicists speak of curving and twisting space-time as if it were a tangible thing. Space-time is little more than a mathematical construct on which to drape the master equations of the cosmos, but an explanation of its origin is nonetheless necessary for physicists to lay claim to a bona fide TOE.

Even string theorists acknowledge this shortcoming, and are in hot pursuit of a "background-independent formulation," one that can account for the cosmic canvas. They might get some help, ironically, from string theory's primary competitor: loop quantum gravity. While it is not as all-encompassing in scope as string theory, the fabric of space-time emerges directly from the equations of loop quantum gravity. The two approaches are quite different in some respects. String theory starts at the quantum level and builds upward to incorporate general relativity, while loop quantum gravity starts out at the top with general relativity and seeks to incorporate quantum mechanics. But both involve some kind of loop—loops of string in string theory, and the mathematical equivalent to loops of space in loop quantum gravity. This suggests that the two camps may, in fact, turn out to be a sort of duality: opposite ends of the same TOE.

Chances are that string theory will not be verified in one dramatic swoop in the experimental equivalent of the portal that opens up behind Fred during her lecture, but in slow, small increments as physicists painstakingly sift through huge amounts of data and stitch together the bits and pieces needed to support specific predictions. It may turn out that string theory, in its present incarnation, fails the test. That doesn't spell its doom any more than the shattered supersymmetry of the original ten-dimensional universe ended its existence, or Fred and Gunn's broken romantic bond ends their underlying friendship. String theory will adapt and continue in a new incarnation, which will in turn be subject to testable predictions, and so on, until physicists arrive at their ultimate goal: a workable theory of everything. That's how science advances. Failure is an acceptable option, even if it's not the most desirable one. There is no progress without it.

Symmetry breaking is a kind of failure that, while traumatic, gives rise to life, the universe, and everything. This might provide a modicum of comfort to Fred and Gunn. Change is built into the mechanisms of the cosmos, right down to the level of kaons and B mesons. Perfection is a static and therefore unnatural state that cannot—and probably should not—be maintained, as Gene the temporal physicist learned when he tried to freeze that perfect moment in time and prevent his girlfriend from breaking up with him. The shattering of perfection, in countless tiny different ways, is what gives meaning to human existence. "The wheel keeps turning. You can't stop it," Lorne tells a rueful Gene, pointing out that while he can hold a musical note indefinitely, "Eventually that's just noise. It's the change we're listening for, the note coming after, and the one after that. That's what makes it music." Lorne's musical analogy is particularly apt for string theory, since it is the changing vibrations of all those tiny strings that create the diverse "music" of the universe.

10

THE PLURAL OF APOCALYPSE

There's More Than One Way to End a Universe

"No man has learned anything rightly until he knows that every day is Doomsday."
—RALPH WALDO EMERSON

Earthquakes in sunny Southern California generally indicate little more than the random shifting of land masses. So when an earthquake rocks Sunnydale in "Doomed" (B-4), hardly anyone, even Giles, assumes that it means anything more than that. Only Buffy fears that the quake might be a portent of yet another impending apocalypse. Buffy's trepidation proves correct. It turns out that a trio of demons is collecting ingredients for a ritualistic sacrifice to once again open the Hellmouth, thereby releasing a horde of demons on an unsuspecting Earth.

Naturally the Scoobies save the day, but not without a strong sense of déjà vu. As Willow protests, "We did this already!" They've done it twice by then, in fact: once when the Master rose ("Prophecy Girl," B-1), and again in "The Zeppo" (B-3), when an apocalyptic cult tried to open the Hellmouth. Sunnydale's Hellmouth provides a handy end-of-the-world plot device whenever things get too quiet. But it isn't the only prospect spelling doom for the human race. The world of the Buffyverse always seems to be on the verge of imminent destruction by some evil force or another, prompting Buffy's college boyfriend, Riley, to moan that whenever he hears her war stories, he finds himself needing to know the plural of apocalypse ("A New Man," B-4).

"The Earth is doomed," Giles declares at the end of "The Harvest" (B-1), in mock despair at Buffy's carefree attitude toward the future battles she faces. He makes a valid point. The threat of apocalypse is no less real in our own world. Our "demons" come in the form of Mother Nature and our own careless self-destructiveness: If a random asteroid doesn't destroy life on Earth, chances are that we'll make our planet uninhabitable through global warming, nuclear warfare, or some equivalent man-made catastrophe. The universe itself will outlast us all, but even that is marching toward its own inevitable death. It's just happening on cosmic time scales that extend far beyond the life span of the human race—even beyond the lifetime of the Earth itself.

DEVOURER OF WORLDS

A construction crew unearths a large stone obelisk carved with ancient runes in "Becoming" (B-2). The relic turns out to be more than just a big decorative rock. It's the tomb of an ancient demon named Acathla, whose mouth serves as a portal to hell. Legend has it that before the demon could quite literally swallow the world, a brave knight pierced his heart with a sword, and Acathla turned to stone. If a demon performs the requisite blood ritual correctly, Acathla will "awaken," opening his mouth to create a sucking vortex that will get bigger and bigger, until he engulfs everything on Earth. It's the yawn to end all yawns.

Ancient demons with the ability to gobble up a universe are, needless to say, a rarity, but there are a couple of similar real-world doomsday scenarios that pop up in theoretical physics. One possible connection is black holes. The vortex created when Acathla opens his mouth is essentially a type of portal that connects Sunnydale to an alternate hell dimension. Ergo, it must share some physics elements with a black hole, with one notable exception: Black holes collapse inward to a point of infinite density (the singularity). They don't grow larger and spread outward like Acathla's gaping maw. If they did, the universe would have ended a long time ago, since a black hole's event horizon—the hypothetical point of no return—would expand along with the object itself. Still, a black hole's strong gravitational pull "swallows" anything that crosses its event horizon, so anything that gets sucked into a black hole is lost, for all practical purposes. What happens after that is anyone's guess—and the subject of much theoretical speculation—but descending into Acathla's gullet, and thence into hell, might seem an attractive option in comparison.

The outcome would indeed be catastrophic if we could create a full-size black hole in the laboratory. Fortunately, this is impossible, at least with our present capabilities. Black holes arise from the immense gravitational pull of a collapsing massive star, which wouldn't fit on our planet, let alone in a laboratory. Yet some physicists have speculated that the next generation of particle accelerators—perhaps even the Large Hadron Collider (LHC)—might be able to produce mini black holes, along with the usual host of subatomic particles. There's just one catch. In order for the LHC to be powerful enough to concentrate sufficient mass in a small enough space to produce mini black holes, string theory must prove to be correct in its assumption that gravity isn't as weak as we think, but merely disperses into extra spatial dimensions. In that case, gravity would be much stronger at the smallest subatomic distances (the Planck length) where the extra dimensions should exist. If gravity were stronger, the LHC would need far less energy than previously expected to achieve the requisite densities for creating mini black holes. Under those rare conditions, the facility would be capable of producing as many as 86,000 every day.

This is admittedly a long shot. Even if they were to form, such mini black holes would be too tiny—about the size of an electron—to pose any serious risk to the macroscale world. They would also be extremely short-lived. Recall from chapter 5 that black holes can emit tiny particles of radiation, causing the black holes to gradually lose mass and eventually evaporate out of existence. Normally this happens on cosmological time scales, but the evaporation rate is dependent on the size of the black hole. The smaller the black hole, the faster it evaporates. Any mini black holes produced by the LHC would be so tiny, they would evaporate in fractions of a second, long before they could make like Acathla and swallow up the Earth.

That hasn't allayed public fears about the potential risks involved in re-creating such volatile conditions in the lab. When Brookhaven National Laboratory prepared to fire up its Relativistic Heavy Ion Collider (RHIC) in 1999, rumors began swirling in the media that the machine could give rise not only to mini black holes,* but also to a cataclysmic chain reaction capable of wiping out the entire universe. The *Sunday Times* of London ran a panicked headline declaring "Big Bang Machine Could Destroy Earth!" while the *Chicago Sun Times* began publishing regular public safety alerts as the RHIC's opening date neared. There were even calls for Congressional hearings and a request that the facility's opening be delayed indefinitely. The absurdity hit an all-time high when Brookhaven officials received a call from a reporter asking—in all seriousness—whether the new collider could have created a black hole that swallowed the plane of John F. Kennedy Jr. as it flew past Martha's Vineyard.

The RHIC has been operating for several years now, and our universe is still intact, so the hysteria proved unfounded. Yet there was a tiny grain of truth to that fear, even if the likelihood of such a disaster occurring is practically nil. The RHIC is designed to simulate the conditions that existed immediately after the big bang. It does this by accelerating gold atoms to near the speed of light. The nuclei collide with so much force, the impact produces enormous energies, and temperatures a million times hotter than the core of our sun. At those temperatures, a new form of matter should emerge called a quark-gluon plasma (QGP), a primordial soup of the quarks and gluons that make up individual protons and neutrons—only in a QGP, they will be able to roam freely instead of being bound tightly into protons and neutrons.† It's a kind of phase transition, such as when ice melts into water, except that the RHIC is melting atoms. In this way scientists hope to learn more about how our early universe formed.

*Some theorists contend that the RHIC has already created a fireball with properties strikingly similar to those of a black hole—although, as expected, it was too short-lived to pose any kind of catastrophic risk.

†The exact nature of this matter is still uncertain. In 2005, RHIC scientists announced evidence not for a QGP, but for a quark-gluon liquid.

That doesn't sound especially apocalyptic, so what was the big deal? There is a tiny probability that a RHIC collision could form a bizarre object called a strangelet. All such collisions produce a mixture of the six quark flavors: up, down, charm, strange, top, and bottom. The greater the quark's mass, the more energy is required to produce it, the more rarely it appears, and the less likely it is to stick around for very long. The RHIC's high energies give rise to more collisions, so it naturally produces more quarks in general, and thus more strange quarks. Most of the time, strange quarks decay within fractions of a second into lower-energy up or down quarks. The fear is that the ultra-high pressures inside the RHIC facility could sustain conditions to the point where strange quarks could stick around long enough to combine with up and down quarks and form a strangelet. Even if this happened, those strangelets would also be unstable and wouldn't last long enough to be dangerous, just like the mini black holes. Furthermore, any strangelet would most likely contain fewer strange quarks than up and down ones, so it would be positively charged and harmless.

Here's where the big "what if" comes in: If there were more strange quarks than up and down ones—an even tinier probability than the strangelet forming in the first place—then the strangelet would have a negative charge. That would make it very dangerous, because it would "eat" all matter it encountered. A negatively charged strangelet would strip away the electrons of any normal atom it bumped into and absorb the exposed nucleus, triggering a devastating chain reaction that would continue until all matter in the universe was converted into strangelets. It's not quite the same thing as Acathla gobbling up the world—unless we surmise that when Acathla awakens, his matter magically converts into negatively charged strangelets that spread outward as Acathla's mouth grows larger and larger, gradually converting all ordinary matter into this strange new "Acathla matter." But in both the fictional and real-world scenarios, the end result is the same: The known universe is consumed.

Statistically, there's no real cause for panic. It is far more likely that a strangelet would have a positive rather than a negative charge. The chances of a stable, positively charged (and therefore harmless) strangelet being produced are, at best, on a par with winning the lottery. The chances that the RHIC would produce a negatively charged strangelet and endanger the planet are even smaller: on a par with someone winning the lottery more than ten times in a row. The odds might not even be that good. The Earth's atmosphere is routinely bombarded by cosmic rays, which slam into it with comparable energies to those produced at the RHIC. This has been happening nonstop for billions of years, but those collisions have yet to "win the lottery" and produce a single strangelet. The RHIC would need to operate continuously for even longer than that before it became likely that a strangelet would form. Even then, it would most likely be the harmless, positively charged variety. As MIT physicist Robert Jaffe memorably told *Newsday*, "It's more likely that a spaceship is going to land in the middle of Texas, and that aliens are going to come out and tell us that the New York Yankees are all aliens."

In the Buffyverse, these numbers are far less comforting, since statistically unlikely events happen all the time, thanks to all the extra mystical energy floating about. That energy is supplied in "Becoming" (B-2) by Angelus's blood. His blood is also the only way to reverse the process. So Buffy runs a sword through Angelus's heart. This sends him through the vortex into hell, thus closing the "portal" and saving the world.

Matters are far less straightforward in the real world. We can't put a sword through the heart of the public's fears and its innate distrust of science. The RHIC's hypothetical strangelet scenario is not the first time a scientific experiment has raised fears of a world-ending catastrophe, nor is it likely to be the last. When physicists began developing the atomic bomb in the 1940s, there were those who worried that a nuclear explosion would set the Earth's atmosphere on fire, while in the 1990s, Fermilab's giant particle accelerator, the Tevatron, generated fears that the facility would create a supernova.

Any number of universe-destroying possibilities will no doubt trigger renewed panic when the LHC goes online in 2008. To high-energy physicists, the LHC is the next logical step in valid scientific inquiry, with the potential to shake the theoretical foundations of the field, yielding revolutionary new insights into nature's most elusive mysteries. But human fear is not logical. To an increasingly suspicious and wary public, that metaphorical shaking could just as easily be viewed as a very literal portent of our own impending doom.

PERMANENT MIDNIGHT

Portents of doom abound in "Apocalypse Nowish" (A-4). It's not just the minor tremors that are shaking buildings all over Los Angeles. Snakes and rats flee the sewers and overrun bathrooms. Swarms of bugs attack random locations, walls begin to bleed, and hundreds of kamikaze sparrows dive-bomb the hotel headquarters of Angel Investigations. The phones are ringing off the hook with panicked reports of unwanted mystical activity.

Everything culminates with the arrival of a big hulking red Beast, complete with horns and cloven hooves. Angel and his cohorts can't stop the creature from punching its fist through a mystical symbol on a rooftop, which produces a towering column of fire that shoots into the night sky and fills the clouds. It's a bit like how moisture evaporates and "seeds" clouds with rain, only in this case, the rain is made of fire. The clouds begin to glow and rain tiny fireballs over the entire city.

A rain of fire is a rare phenomenon, even in the Buffyverse. But falling meteors, also known as shooting stars, could provide a similar eye-popping effect. Meteors are the bits and pieces left in the wake of comets, or chunks that have broken off asteroids. Most burn up from friction with the air as they enter the Earth's atmosphere, long before they reach the surface, although on rare occasions a large meteor won't burn up completely and its remnants will crash into the Earth. They leave a bright trail of light as they streak across the sky; the color of that light depends on how fast they travel, and how hot they burn. The rain of fire that hits Los Angeles is reddish-orange in color, indicating meteors that travel at the lower end of the speed range, around 25,000 mph.*

Alas, the city's troubles are just beginning. In "Long Day's Journey" (A-4), the Beast begins systematically executing five ancient beings known as the Ra-Tet. By killing the Ra-Tet, absorbing their power, and collecting the talismans they carry, the Beast can perform a ritual that blots out the sun, plunging Los Angeles into a permanent midnight that will eventually spread to cover the entire world. He manages to do just that. A dark cloud blocks the sun's rays over Los Angeles.

The city's predicament bears a striking resemblance to the hypothetical disaster scenario of nuclear winter that first emerged in the 1980s. Carl Sagan and a few colleagues crunched some numbers and determined that the detonation of about 400 nuclear weapons in a global nuclear war would ignite huge fires, especially

*Meteors can travel up to 160,000 mph, but at those speeds they appear in the blue to white region of the spectrum.

in heavily populated cities like Los Angeles. This would emit sufficient dust and debris into the Earth's upper atmosphere to block the sun's rays from reaching the Earth's surface for many months. Four hundred bombs might sound like a lot, but it's a fraction of the estimated 24,000 to 33,000 nuclear weapons stored in arsenals around the world, the bulk of them in the United States and Russia.

Apart from the obvious health consequences of lingering radiation and nuclear fallout, all that smoke and soot blocking the sunlight could have disastrous consequences for any members of the human race who survived the initial blast. Since the sun's rays couldn't penetrate the smog, surface temperatures would cool significantly, killing most plant and animal life on Earth. Humans in turn would slowly starve to death. Of course, this is a worst-case scenario, and a somewhat controversial one at that. The surface cooling could be far less severe, resulting in a "nuclear fall" rather than winter. Alternatively, the explosions could produce so many nitrogen oxides that they would damage the ozone layer in the stratosphere, allowing more ultraviolet radiation to reach the Earth's surface, increasing global warming and the greenhouse effect. The one thing scientists can fully agree on is that such a global-scale detonation of nuclear weapons would have a devastating effect on the Earth's climate.

In *Angel*'s Los Angeles, the biggest threat arises from the city being overrun by vampires and other nocturnal hunters, not by the onset of a nuclear winter. We don't see anyone reaching for a sweater to keep warm against the sudden chill, although that might be because the surface cooling effect hasn't had time to kick in just yet. But technically, the sun is still there, and still generating heat—it's just not visibly shining in the greater Los Angeles area. Had the Beast managed to obliterate the sun completely—

perhaps by causing it to explode like a fiery supernova—the effects would have been far more catastrophic, and not so localized.

The people of Los Angeles (and everyone else on Earth, for that matter) wouldn't notice anything amiss for roughly eight minutes. That's how long it takes for the sun's light to travel 93 million miles or so to reach the Earth. Then everyone would be plunged into sudden darkness. And that wouldn't be the worst of it. The sun's mass causes space-time to curve, giving rise to the gravitational force that holds the planets in their orbits. If the sun were obliterated, the sudden absence of mass would cause space-time to spring back into a flat configuration, and that gravitational pull would disappear. What happens then? One possibility is that the Earth would spin off its axis and shoot out of its orbit into deep space, along with all the other planets in the solar system. A more likely scenario is that the Earth would be vaporized by the burst of energy from the sun's supernova explosion long before it was knocked out of orbit.

Fortunately, our sun will end not with a bang but with a whimper, when it finally burns through all the fuel in its core, roughly five to ten billion years from now. Like all stars, the sun is a giant nuclear reactor, made up of hydrogen and helium gas. To produce light, it converts hydrogen into helium through 10^{38} fusion reactions each second. When all the hydrogen is turned into helium, the sun will no longer generate sufficient energy to counteract the pull of gravity. The core will contract inward, heating up the interior to sufficient temperatures to fuse helium into carbon. This would still spell doom for planet Earth and all of its inhabitants. Just before this new phase begins, the sun will puff outward and turn into a red giant, expanding its outer layers as the core collapses. Those outer layers will engulf the nearest planets, including Earth, which will plunge into the core and be vaporized. The sun's own lifetime will continue, however, for a few billion years more before it ultimately cools into a black dwarf star.

KEY TO THE KINGDOMS

Permanent midnight might be preferable to total interdimensional chaos. Buffy mysteriously acquires a younger sister in Season 5 of *Buffy*, who turns out to be a mystical "Key." For centuries, we are told, the Key was just a ball of mystical energy. But when the exiled hell god, Glory, sets her sights on obtaining it, a brotherhood of monks hides the Key in human form and sends it to Buffy for protection ("Blood Ties," B-5). As her "Key" moniker implies, Dawn opens an interdimensional lock. Glory just wants to return to her hell dimension, but to do so, she must open a portal.

Portals open all the time in the Buffyverse, but this time the situation is different. The Key doesn't open just one portal, but all gateways to all dimensions simultaneously ("Spiral," B-5). Once the Key is ritualistically activated, "The walls separating realities will crumble, dimensions will bleed into each other, and the universe will tumble into chaos, disorder, and ultimate darkness." So says the Byzantine general, the doom-and-gloom leader of a strange sect intent on destroying the Key. Dawn could bring about the ultimate apocalypse, one that affects every species in every dimensional world in the Buffyverse.

The means by which the Key accomplishes this impressive feat—aside from the usual ritualistic bloodletting—isn't entirely clear, but we can ferret out clues from the above-cited episodes to build a case for a possible underlying mechanism. Most notably, the Key is described as "an energy matrix vibrating at a dimensional frequency beyond human perception." This tells us that parallel dimensions in the Buffyverse possess separate frequencies that serve as natural energy barriers, or "walls," between themselves, much like radio stations differentiate themselves by broadcasting at specific radio wavelengths. Usually, receivers only pick up those transmitted signals to which they are precisely tuned. Dawn seems to be a kind of mystical transmitter that is tuned not just to one specific frequency, but to all possible frequencies at once. The "signal" she transmits is the equivalent of interdimensional white noise.

We also learn that Dawn's universe-destroying potential can only be realized if she is activated at a specific time and place ("The Weight of the World," B-5). All the dimensional frequencies must be precisely aligned before their respective portals can open simultaneously. The notion calls to mind forced oscillation resonance (described in chapter 1), where Buffy's prolonged scream so precisely matches the natural resonant frequency of the Gentlemen's heads that it causes them to explode. The Key matches all resonant frequencies, so when it is activated at the precise time and place, its energy causes all the dimensions to vibrate together. The combined energy builds up until the dimensional walls begin to crumble. In this case, it is not demonic heads that are exploding, but the very fabric of space-time that is coming apart at the seams.

What might be the origin of these bad vibrations? Since the Key is described in terms of energy and vibrations, and opens all portals, one's thoughts naturally turn to string theory for more insight into the physics that might underlie this particular scenario. After all, string theory is the best candidate we have so far for a "master key" to unlock the secrets of our universe. Recall that everything we see around us may have a teensy oscillating string at its center. It is the different ways in which the tiny strings vibrate that give rise to all subatomic particles and fundamental forces—not to mention the fictional "dimensional frequencies" separating the other worlds that populate the Buffyverse. All strings possess a fundamental property called tension; they are stretchy, like rubber bands. Just as with guitar strings, the more a string is stretched, the higher the tension, and the more vibrating energy that string will possess.

Because strings are so elastic, space-time is much more changeable and dynamic in string theory than in relativity or quantum mechanics. It can curve in response to the presence of mass or energy, and it can compactify to such enormous densities that the fabric tears, then repairs itself and reinflates into any number of possible configurations. Strings compactify and reinflate along with space-time, providing a protective "shrink-wrapping" around any tear that may occur. This prevents that tear from ripping out of control. And if space-time can tear and reconfigure (as we learned from Fred's fictional lecture in "Supersymmetry," A-4), then wormholes, or portals, to other regions of space-time—perhaps even to parallel worlds—may be possible. We've seen that space-time must rip in order for a single portal to open. For all portals to open at once, space-time would have to be shredded. When the Byzantine general speaks of crumbling dimensional walls in "Spiral" (B-5), he is describing, in essence, a series of rips in the fabric of space-time without the protective shrink-wrapping. The Key gives rise to the dreaded "naked singularity"—not just once, but many times over.

In such a hypothetical scenario, a small tear forms in the fabric of space-time, but without its usual safeguard, so scientists would be able to observe it directly. Physicists have debated for decades whether naked singularities are possible. If they do exist, many believe that an innate "cosmic censorship" would shield them from direct observation (and conveniently prevent them from spreading out of control). The most notable naysayer is Stephen Hawking, who famously made a bet with two CalTech physicists in 1991 that naked singularities could not exist, although the stakes weren't especially high: 100 pounds sterling, plus an article of clothing "embroidered with a suitable concessionary message."

Hawking was forced to backpedal a bit on his stance in 1997, after new calculations revealed a limited set of conditions that could give rise to naked singularities. Ironically, it was his own work on black holes that provided those conditions. Most physicists agree that singularities lie at the center of black holes, but these are "clothed," conveniently hidden from view beyond the event horizon, from whence nothing—not even light—can escape, thanks to the black holes' immense gravitational field. Yet Hawking showed that black holes radiate energy and decay over time. Assuming that the protective event horizon evaporates along with the black hole, the singularity at the center could be exposed at the moment the black hole winks out of existence. It's highly improbable, but even Hawking admits that it is theoretically (i.e., mathematically) possible.

He may have lost the bet on a technicality, but Hawking still adheres to a general cosmic censorship principle, and the message on the T-shirt he presented was far from concessionary: "Nature abhors a naked singularity." Small wonder that Nature is such a prude, considering the potentially catastrophic effects if a naked singularity were allowed to tear unchecked. That's the kind of apocalyptic scenario the Scoobies face when Dawn's blood activates all the mystical interdimensional portals at once. In the Buffyverse, it seems that cosmic censorship can be mystically overcome. The Key's vibrational energy (triggered by the flow of Dawn's blood) causes the energy walls of all dimensions to vibrate simultaneously in response. Those vibrations become stronger and stronger the longer the Key is active, weakening the energy barriers between dimensions until they rip. The dimensions begin to bleed into each other, unchecked.

Fortunately, the Buffyverse also provides a handy means of repairing this cataclysmic rip. When the Key's incoming mystical energy stops—when Dawn is dead and her blood no longer flows—the resonant vibrations of all the dimensional walls will cease, the stringy safeguards will snap back into place, and space-time will heal itself. So the faster Dawn dies, the better the Buffyverse fares. Hence the mantra of the Knights of Byzantium: "The Key is the link. The link must be severed." At the critical moment, Buffy realizes that they share the same blood. Rather than let her sister die, she dives into the growing vortex, and is killed en route to the bottom by the powerful mystical energies that created it ("The Gift," B-5). This closes the rip, thwarting the fledgling apocalypse. Once the critical moment of dimensional alignment has passed, Dawn becomes a harmless teenager.

TIPPING THE SCALES

A fundamental cosmic imbalance is the cause of the final near-apocalypse faced by Buffy and her gang of Scoobies in Sunnydale. In this case, it's not a literal catastrophic rip in space-time, but a metaphorical one: an attempt on the part of the First to exploit a tiny vulnerability until it widens, with the sole aim of wiping out the entire Slayer line. In "Showtime" (B-7), we learn that the Slayer line has become vulnerable because the balance of the Buffyverse has been disturbed by Buffy's second resurrection. She has

$Z = \int D^F[\rho(\xi)] \exp\left(\frac{(26-D)}{12\pi}\int_\xi\left[\frac{1}{2}\frac{(\partial_a\rho)^2}{\rho^2}\right] + \int_\xi \mu_R^2\rho^2\right)$
"THE BLOOD FLOWS, THE GATES
OPEN."

violated its built-in laws of conservation. Balance must be restored, lest evil overrun the world unchecked.

A brief review of the origins of the Slayer, and the mechanism by which the Slayer essence is handed down to progressive generations, is in order here. We learn in "Get It Done" (B-7) that a tribe of primitive Shadow Men created the very first Slayer. They chained a young girl to a rock and forced her to mate with the essence of a demon, thereby absorbing its extra mystical strength. Because it came from the same dark source, the Slayer's power matched that of the demons, enabling her to fight them more effectively. But there can be only one Slayer at time. When one Slayer dies, another takes her place, inheriting the mystical powers. At any given time, there are numerous Potentials—young girls with certain innate gifts that make them natural candidates to be a Slayer. They are waiting to be activated with that extra influx of energy that gives the Slayer her unique advantages—a handy metaphor for the conversion of potential into kinetic energy.

Traditionally, when Slayers die, they stay dead. Buffy is not so much with the tradition. First, she drowns in a confrontation with the Master ("Prophecy Girl," B-1). Xander revives her, but she was clinically dead long enough for a second Slayer, Kendra, to be activated ("What's My Line," B-2). As narrative devices go, this unexpected anomaly provided the writers with a rich motherlode of themes and plot twists to mine in subsequent episodes, but scientifically, it's problematic. Having two Slayers at the same time technically violates the conservation laws of the Buffyverse (outlined in chapter 3). All Buffy's power should have been transferred to Kendra when Buffy drowned. Death is usually an irreversible process, so the power is transferred in one direction only, from the dying Slayer to a Potential. By coming back from the dead, Buffy reverses the irreversible. There are now two Slayers, but if energy is conserved, the total amount of Slayer energy available remains constant. So all that power must be split between Buffy and Kendra, making each half as strong. Either that, or one of them must be "unchosen" and forfeit her powers. Yet this is clearly not the case. Buffy appears to have lost none of her considerable powers, and Kendra easily matches her. In the absence of any evidence of an extra influx of Slayer energy coming from an outside source, we must conclude that energy conservation has been violated, upsetting the balance of the Buffyverse.

The system gamely tries to correct itself via a built-in safety mechanism. When Kendra dies at Drusilla's hand ("Becoming," B-2), Faith is activated ("Faith, Hope, and Trick," B-3), yet when Buffy dies again, another Slayer is not activated. The writers have explained in interviews that each Slayer gets only one "replacement": She can activate only one Potential with her death, and Buffy has already done so. Faith must die before another Slayer can be activated. It is the Buffyverse's way of ensuring that energy is once again conserved after a brief violation.* The principle at work here is a bit like the way virtual particles can momentarily violate energy conservation as long as they can't be directly observed. They don't exist long enough to permanently upset the balance of the universe. The first time Buffy died, it was a technicality: "I was only gone for a minute," she says defensively upon discovering Kendra's existence. Buffy's second demise should have been permanent, restoring the balance.

Unfortunately, Buffy comes back again three months later, literally clawing her way out from her own grave in response to Willow's resurrection spell. What ought to have been a fluke, a temporary imbalance, becomes firmly entrenched. This is where

*The four-season interval between Buffy's first and second demise might not seem all that brief to TV viewers, but it is just a blip in history from the perspective of a timeless entity like the First.

the First sees its opening—and where the writers of the series apparently saw their chance to address the violation problem directly. One by one, agents of the First begin taking out Potentials. The reasoning is simple: If there are no more Potentials, a new Slayer can't be activated, and Faith's death will mark the end of the Slayer line ("Bring on the Night," B-7). All the First needs to do then is take out Buffy—easier said than done—and hope that this time she doesn't rise again.

Even before Buffy threw everything out of whack by refusing to stay dead (twice), it could be argued that the Slayer line must be weakening over time. Entropy dictates that every time the primordial power is transferred to a new Slayer, a little bit of energy must be lost as heat in the transformation, because no energy transfer can be 100 percent efficient. So the Slayer line ought to be getting weaker with each successive incarnation, unless it is not a "closed system": That is, there must be a corresponding influx of energy coming from another source to make up for the inevitable energy loss, much like a refrigerator can repeat its cooling cycle indefinitely only if it is plugged into an electrical outlet, thereby drawing enough extra energy to offset the effects of entropy. All the talk about balance is a bit overly optimistic, even if there is such an energy source feeding into the Slayer line. Pitting one lone Slayer against all the hordes of hell was grossly unfair from the outset. The odds are definitely in evil's favor, even more so if one factors in entropy.

Facing imminent annihilation from the First's army of *über*-vamps, Buffy is offered more power by the mythical Shadow Men in "Get It Done" (B-7). She refuses, because it would require her to merge with the essence of a demon, just like the very first Slayer, and any sensible young woman would balk at that prospect. Still, the experience alerts her to the need for bringing a much-needed infusion of mystical energy into the depleted Slayer line. That's exactly what Buffy decides to do in "Chosen" (B-7), only this time, the infusion stems from a less-tainted source. She taps into the power of a mystical scythe bequeathed to the Slayer line by an ancient group of women known as the Guardians. Willow finds a way to release this new kind of energy, so that instead of only one Slayer being activated at a time, all the Potentials are activated at once, producing an instant army of Slayers. In terms of the magnitude of its impact, the effect is a bit like the difference between the typical decay of an atomic nucleus, which releases energy gradually, and the induced nuclear chain reaction that powers an atomic bomb.

And it does the trick. Thanks to this massive infusion of energy, Buffy and her new army of Slayers prevail, beating back the hordes of hell and sending the First into humiliating retreat. Their victory might not be decisive, in that the First can't really be killed. But it does help even the odds, reversing some of the inroads entropy has made over the centuries, not just in the Slayer line, but in the ongoing battle between good and evil. This isn't an option in the real world. Entropy has proven thus far to be an inviolable law of physics. We have no magic scythe or similar mystical source of extra energy on which to draw to offset entropy's effects in the universe at large, which (to the best of our empirical knowledge) appears to be a closed system. That's one reason the second law of thermodynamics may very well prove to be a downfall of sorts for our own universe.

WHY WE FIGHT

While Sunnydale is beset by a series of separate apocalyptic scenarios, the notion of one all-encompassing Apocalypse (with a capital "A") of biblical proportions weaves through all five seasons of *Angel*, with the soulful vampire supposedly destined to play a critical role in tipping the scales—for good, or evil, no prophecy can say. Team Angel naturally assumes that the coming Apocalypse will be a single, decisive, cataclysmic event, an all-out war between good and evil that will decide the fate of the human race. They receive a rude awakening when they learn from Wolfram & Hart's former wunderkind lawyer, Lindsey McDonald, in "Underneath" (A-5) that the Apocalypse is already well under way. In fact, Lindsey says with a smirk, "You're soaking in it." Far from being a single event, it is a slow meandering hellward, spanning years, perhaps even centuries.

Give evil credit for adopting an ingenious strategy. A sudden cataclysm would draw too much attention. People would rise up and fight, if for no other reason than to save their own skins. This way, humanity is anesthetized into complacency, learning to accept the world the way it is, so that no one notices the inexorable slide "toward entropy and degradation," as Lindsey so evocatively puts it. Even Team Angel has bought into the lie. Taking charge of Wolfram & Hart's Los Angeles branch has kept it distracted from its mission of fighting the world's downward descent.

It's interesting that Lindsey mentions entropy, because the second law of thermodynamics provided the basis for a nineteenth-century theory about how the universe will end—another slow-moving apocalypse, this time on cosmic time scales. Rudolf Clausius and William Thomson, Lord Kelvin, were among those scientists who believed in the "heat death" of the universe. The term is a bit of a misnomer, since the theory holds that the universe will end in ice: Essentially, it will freeze to death as a result of entropy. Recall that all physical systems tend toward a state of equilibrium, in which there is no net flow of energy. For instance, gas molecules are confined by the walls of their container, but when the gas is released, the molecules disperse outward until they are evenly spread out into the larger space. Clausius and Kelvin reasoned that if the second law were extrapolated to its logical conclusion, the entire universe would eventually reach a similar state of equilibrium, with a uniform temperature throughout its vast expanse. Change would stop, and therefore time would cease. The stars would gradually run out of fuel, plunging the universe into an eternal darkness far more terrifying than the Beast blotting out the sun. Needless to say, all life would come to an end.

The eventual death of the cosmos might be a slow meandering from our mortal human perspective, but the outward expansion of the universe is actually moving progressively faster. Physicists attribute this unexpected acceleration to the presence of a mysterious form of repulsive gravity (antigravity), which they have dubbed "dark energy." This has some intriguing implications for the ultimate fate of the universe. Among the more fantastical theories is that the dark energy is so repulsive that it will ultimately rip apart all the galaxies, stars, and planets in the cosmos—even the atoms themselves. Dubbed the "Big Rip," the theory assumes that the rate of cosmic acceleration will increase instead of proceeding at a constant rate.

It's a subtle, yet critical, distinction. The Scoobies flee Sunnydale in an abandoned school bus after the final battle against the First's army of *über*vamps, barely escaping the city limits before the town collapses into a gigantic crater. (So much for Buffy's planned post-apocalyptic shopping spree at the Sunnydale mall.) Let's say the school bus represents our universe. If the bus moves 10 mph faster with each mile it travels, it accelerates at a constant rate. But if the school bus accelerates an extra 10 mph every mile, then every half mile, every hundred yards, and so on, it accelerates at an ever-increasing rate, until the bus begins to break apart from the sheer force of that acceleration.* That's what would happen to the cosmos in the Big Rip scenario: The universe would experience the g forces from hell. Eventually (in roughly 20 billion years), the rate of expansion would reach such a phenomenal pace that gravity would be overwhelmed completely, along with the nuclear forces that bind subatomic particles to form matter. Everything in the cosmos would break apart, right down to individual atoms.

The Big Rip is not the most likely scenario, however. Based on all the experimental evidence to date, it appears that the rate of acceleration is constant,† indicating that the expansion of the universe will continue indefinitely at a fixed rate. This means that Clausius and Kelvin may have been right after all: Our universe most likely will end in ice. All matter in the universe will grow farther and farther apart, until hardly any galaxies or stars will be visible from our tiny corner of the Milky Way. In fact, some physicists estimate that, although there are roughly 100 billion galaxies in the universe, within 150 billion years only a few thousand will be visible from Earth—not that we'll notice. By that time, our sun will have long ago burned out, ending life as we know it, and possibly consuming the Earth itself. The universe will continue expanding until it does indeed experience a heat death. The

*These are the g forces. A g is a unit for measuring acceleration in terms of gravity. The two are equivalent, as we saw in chapter 5. G forces arise whenever something accelerates (as opposed to standing still or moving at a constant speed), because of the corresponding variations in the strength of gravity's pull. The faster something accelerates, the stronger the g forces.

†At the January 2006 meeting of the American Astronomical Society, one astronomer begged to differ, presenting data on distant gamma ray bursts as preliminary evidence that the rate of acceleration may indeed be speeding up. Those results await further testing. Prior measurements of cosmic expansion were based on supernova explosions, which most scientists still believe to be the most accurate indicator.

temperature of deep space will plunge ever closer to absolute zero, until molecules can barely move. Even black holes will evaporate away into energy, leaving nothing behind but drifting subatomic particles.

We will all be long gone before that happens, of course, which dampens any strong sense of apocalyptic urgency (hence the brilliance of Wolfram & Hart's snail-paced strategy). What could we do about it anyway? Our options are limited, but Michio Kaku has speculated that a highly advanced future civilization may be able to fashion an escape hatch of sorts in order to survive the increasingly hostile conditions. For instance, it might one day be possible to give birth to our very own baby universe in the lab, although this would require concentrating incredibly intense energy in a highly pressurized chamber to attain scorching temperatures of around 1,000 trillion degrees—hotter than the interiors of the hottest stars. Only the moment of the big bang itself boasted energies of that magnitude. That is the critical "boiling point" when tiny bubble universes would begin to form, connected by wormholes. Most would be very short-lived, but at such high pressures and temperatures, a few may last long enough to inflate into full-fledged baby universes.

If the connecting wormhole between our universe and the newborn world could be made traversable—no small feat, as we saw in chapter 9—an advanced civilization could conceivably use the wormhole to escape the dying cosmos, building a new life in the fledgling baby universe. And if a traversable macroscopic wormhole still proves impossible, even to an advanced civilization, Kaku speculates that it may be possible to condense the information needed to re-create such a civilization down to subatomic scales, and then "inject" this "seed" through a tiny wormhole into a baby universe, where it will self-assemble on the other side.

Hypothetical speculation is one thing; practical implementation is quite another. Respected scientist though he is, what Kaku is proposing remains relegated to the farthest reaches of science fiction for the foreseeable future. But science fiction is the natural habitat of the Buffyverse, which isn't limited by such trifling matters as insufficient energies, radiation feedback loops, or technological impossibilities. In fact, a similar approach to Kaku's inventive strategy appears to have been adopted by the Old Ones. The series mythology holds that in ancient times, the pure demon races left the human dimension in a mass exodus when it became clear that the humans had overrun the planet. They must have left by way of a portal (a wormhole) and ventured into a new dimension where they could live free of the human pestilence.

Since then, the tide has shifted in favor of the remaining "mixed-race" demons. Should conditions continue to deteriorate, there's no reason any few surviving humans in the Buffyverse couldn't follow suit, using a portal to set up shop in some parallel dimension when retreat is the only remaining option. That day is arguably inevitable. Remember C. P. Snow's summation of the laws of thermodynamics: We can't win, we can't break even, and we can't get out of the game. The same basic "house rules" apply in the Buffyverse, although it has a few more exotic loopholes at its disposal than we do back in the real world. *Buffy* ends on an optimistic note, with Buffy finding a way to reverse the cumulative effects of entropy on the depleted Slayer line, at least temporarily, by bringing in extra mystical energy. *Angel*'s final outlook is quite a bit bleaker, preferring to face the harsh truth head-on: Sometimes our best efforts aren't enough. Entropy is a war we are destined to lose, even if we occasionally win an individual battle.

That doesn't mean the war isn't worth waging. As Lindsey reminds us, "Heroes don't accept the world the way it is; they fight it." That's why Angel decides to infiltrate the Circle of the Black Thorn, a secret cabal of powerful beings who serve as the senior partners' representatives on the mortal coil ("Power Play," A-5). "We're in a machine and the Black Thorn runs it," he tells the skeptical members of his team. "That machine is gonna be here long after our bodies are dust." But by killing the current Circle members, Angel reasons, they can bring the gears to a grinding halt, just for a moment. It is at best a Pyrrhic victory. In the cliffhanger series finale ("Not Fade Away," A-5), a teeming horde of every breed of demon imaginable—including a giant flying dragon—converges on the surviving members of Team Angel. Facing lamentably impossible odds, Spike asks Angel to be a bit more specific about coordinating a plan of attack. "Well, personally, I kind of want to slay the dragon," Angel replies. And the series fades to black.

Our universe is also a gigantic machine. It's been running for some 13 billion years, and it will still be running long after we turn to dust. Yet physicists are not content to blindly accept the mere fact of its existence, nor are they cowed by their own seeming insignificance as tiny cogs in the cosmic scheme of things. Instead, they are passionately engaged in an ongoing quest to unlock the most elusive mysteries of nature. Theirs might not be an epic mythological struggle between good and evil, as in the Buffyverse, but the story of scientific inquiry and discovery throughout human history is every bit as grand in scope, ranging from the farthest reaches of the cosmos to the smallest subatomic particle (or string).

Physicists pursue knowledge for the sake of knowledge, evincing a single-mindedness and selfless idealism that not everyone understands—much like Hamilton, Angel's liaison to the senior partners, can't fathom why Angel would continue to fight when there is no longer any personal gain for him. "People who don't care about anything will never understand the people who do," Angel says, a tad self-righteously, to which Hamilton snarkily replies, "Yeah, but we won't care." When it comes to science, we should care, because the big questions that engage physicists have direct bearing on our own origins and ultimate fate. It's all part of man's search for meaning.

Both Lindsey and Hamilton point out that Team Angel can't possibly win a decisive victory over the senior partners. "You don't stand a chance. We are legion. We are forever," Hamilton gloats. Metaphorically speaking, physicists face a similar dilemma. Physics can never lay claim to definitive proof of even its most successful theories. It is the nature of the scientific method that nothing is ever 100 percent proven, merely supported by an ever-growing body of evidence—hence the widespread confusion among nonscientists as to what, exactly, is meant by a scientific theory. There are few certainties, and even fewer moral absolutes, in the world of science, but that doesn't mean the theories aren't true. Those theories are the best explanations we have thus far for observable physical phenomena, and they are based on hard experimental data. It's just that even when physicists amass truly staggering amounts of data, or accurately measure something to, say, the tenth decimal place, there is always the tiniest chance that a theory could fail unexpectedly at the eleventh decimal place, or the twelfth, and so on, in ever-smaller increments. They must compile more and more data to support their theories, refine them in light of new evidence, and test their results again and again, with ever-greater accuracy—just in case. It's a "war" they can never really "win."

Winning isn't the point of science. Angel observes that maybe Wolfram & Hart and the hordes of hell aren't there in order to be beaten, merely to be fought: "Maybe fighting them is what makes human beings so remarkably strong." The same could be said for physicists. The more we discover about our universe, the more we realize how little we know, because each answer seems to lead to a crop of new questions. The quest for knowledge never ends, and that's what makes the struggle so poignant. But perhaps we aren't meant to "beat the cosmic system" by unlocking all its secrets; we just need to ask the questions.

By and large, physicists are comfortable with not having all the answers. In fact, most would say that the presence of so many unanswered questions and controversies in physics is what makes the field exciting: It's a nonstop intellectual adventure. Occasionally, if we are very lucky, we might uncover some gleaming nugget of truth. But for physicists—indeed, for most scientists—it is the search itself, the process of discovery, that matters, and the bigger the challenge, the more compelled they are to seek out new and innovative paths to deeper knowledge about our world. This makes them every bit as heroic as Buffy or Angel, despite being largely unsung champions. They, too, desire to slay the dragon.

NOW LEAVING
THE BUFFYVERSE
NOTICE:
NO VAMPIRES
OR OTHER
PARANORMALS
BEYOND
THIS POINT

BIBLIOGRAPHY

Author's Note: For the sake of brevity, I have listed only those episodes of *Buffy* and *Angel* with elements that were specifically used in the text. But my conclusions were gleaned from viewing both series in their entirety. All episodes produced by Twentieth Century Fox.

BUFFY THE VAMPIRE SLAYER

Season 1

"Welcome to the Hellmouth." Dir. Charles Martin Smith. Written by Joss Whedon. Air date: 3/3/97.

"The Harvest." Dir. John Kretchmer. Written by Joss Whedon. Air date: 3/10/97.

"Witch." Dir. Stephen Cragg. Written by Dana Reston. Air date: 3/17/97.

"Teacher's Pet." Dir. Bruce Seth Green. Written by David Greenwalt. Air date: 3/24/97.

"Angel." Dir. Scott Brazil. Written by David Greenwalt. Air date: 4/14/97.

"I Robot . . . You Jane." Dir. Stephen Posey. Written by Ashley Gable and Thomas Swyden. Air date: 4/28/97.

"The Puppet Show." Dir. Ellen Pressman. Written by Dean Batali and Rob Des Hotel. Air date: 5/5/97.

"Nightmares." Dir. Bruce Seth Green. Teleplay by David Greenwalt. Story by Joss Whedon. Air date: 5/12/97.

"Out of Mind, Out of Sight." Dir. Reyza Badiyi. Teleplay by Ashley Gable and Thomas Swyden. Story by Joss Whedon. Air date: 5/19/97.

"Prophecy Girl." Dir. Joss Whedon. Written by Joss Whedon. Air date: 6/2/97.

Season 2

"School Hard." Dir. John Kretchmer. Teleplay by David Greenwalt. Story by Joss Whedon and David Greenwalt. Air date: 9/29/97.

"Inca Mummy Girl." Dir. Ellen Pressman. Written by Matt Kiene and Joe Reinkemeyer. Air date: 10/6/97.

"Reptile Boy." Dir. David Greenwalt. Written by David Greenwalt. Air date: 10/13/97.

"Halloween." Dir. Bruce Seth Green. Written by Carl Ellsworth. Air date: 10/27/97.

"What's My Line?, Part I." Dir. David Solomon. Written by Howard Gordon and Marti Noxon. Air date: 11/17/97.

"What's My Line?, Part II." Dir. David Semel. Written by Marti Noxon. Air date: 11/24/97.

"Ted." Dir. Bruce Seth Green. Written by David Greenwalt and Joss Whedon. Air date: 12/8/97.

"Innocence." Dir. Joss Whedon. Written by Joss Whedon. Air date: 1/20/98.

"Phases." Dir. Bruce Seth Green. Written by Rob Des Hotel and Dean Batali. Air date: 1/27/98.

"Killed by Death." Dir. Deran Serafian. Written by Rob Des Hotel and Dean Batali. Air date: 3/3/98.

"Becoming, Parts I and II." Dir. Joss Whedon. Written by Joss Whedon. Air dates: 5/12/98 and 5/19/98.

Season 3

"Anne." Dir. Joss Whedon. Written by Joss Whedon. Air date: 9/29/98.

"Dead Man's Party." Dir. James Whitmore, Jr. Written by Marti Noxon. Air date: 10/6/98.

"Faith, Hope, and Trick." Dir. James Contner. Written by David Greenwalt. Air date: 10/13/98.

"Revelations." Dir. James Contner. Written by Douglas Petrie. Air date: 11/17/98.

"The Wish." Dir. David Greenwalt. Written by Marti Noxon. Air date: 12/8/98.

"Amends." Dir. Joss Whedon. Written by Joss Whedon. Air date: 12/15/98.

"Gingerbread." Dir. James Whitmore, Jr. Teleplay by Jane Espenson. Story by Thania St. John and Jane Espenson. Air date: 1/12/99.

"Helpless." Dir. James Contner. Written by David Fury. Air date: 1/19/99.

"The Zeppo." Dir. James Whitmore, Jr. Written by Dan Vebber. Air date: 1/26/99.

"Bad Girls." Dir. Michael Lange. Written by Douglas Petrie. Air date: 2/9/99.

"Doppelgangland." Dir. Joss Whedon. Written by Joss Whedon. Air date: 2/23/99.

"Choices." Dir. James Contner. Written by David Fury. Air date: 5/4/99.

"Graduation Day, Parts I and II." Dir. Joss Whedon. Written by Joss Whedon. Air dates: 5/18/99 and 7/13/99.

Season 4

"Living Conditions." Dir. David Grossman. Written by Marti Noxon. Air date: 10/12/99.

"Fear Itself." Dir. Tucker Gates. Written by David Fury. Air date: 10/26/99.

"The Initiative." Dir. James Contner. Written by Douglas Petrie. Air date: 11/16/99.

"Pangs." Dir. Michael Lange. Written by Jane Espenson. Air date: 11/23/99.

"Hush." Dir. Joss Whedon. Written by Joss Whedon. Air date: 12/14/99.

"Doomed." Dir. James Contner. Written by Marti Noxon, David Fury, and Jane Espenson. Air date: 1/18/00.

"A New Man." Dir. Michael Gershman. Written by Jane Espenson. Air date: 1/25/00.

"The 'I' in Team." Dir. James Contner. Written by David Fury. Air date: 2/8/00.

"Good-bye Iowa." Dir. David Solomon. Written by Marti Noxon. Air date: 2/15/00.

"Superstar." Dir. David Grossman. Written by Jane Espenson. Air date: 4/4/00.

"Where the Wild Things Are." Dir. David Solomon. Written by Tracey Forbes. Air date: 4/25/00.

"Primeval." Dir. James Contner. Written by David Fury. Air date: 5/16/00.

Season 5

"Buffy vs. Dracula." Dir. David Solomon. Written by Marti Noxon. Air date: 9/26/00.

"The Replacement." Dir. James Contner. Written by Jane Espenson. Air date: 10/10/00.

"Out of My Mind." Dir. David Grossman. Written by Rebecca Rand Kirshner. Air date: 10/17/00.

"No Place Like Home." Dir. David Solomon. Written by Douglas Petrie. Air date: 10/24/00.

"Fool for Love." Dir. Nick Marck. Written by Douglas Petrie. Air date: 11/14/00.

"Shadow." Dir. Daniel Attias. Written by David Fury. Air date: 11/21/00.

"Listening to Fear." Dir. David Solomon. Written by Rebecca Rand Kirshner. Air date: 11/28/00.

"Triangle." Dir. Christopher Hibler. Written by Jane Espenson. Air date: 1/9/01.

"Blood Ties." Dir. Michael Gershman. Written by Steven DeKnight. Air date: 2/6/01.

"Crush." Dir. Daniel Attias. Written by David Fury. Air date: 2/13/01.

"I Was Made to Love You." Dir. James Contner. Written by Jane Espenson. Air date: 2/20/01.

"Forever." Dir. Marti Noxon. Written by Marti Noxon. Air date: 4/17/01.

"Intervention." Dir. Michael Gershman. Written by Jane Espenson. Air date: 4/24/01.

"Spiral." Dir. James Contner. Written by Steven DeKnight. Air date: 5/8/01.

"The Weight of the World." Dir. David Solomon. Written by Douglas Petrie. Air date: 5/15/01.

"The Gift." Dir. Joss Whedon. Written by Joss Whedon. Air date: 5/22/01.

Season 6

"Bargaining." Dir. David Grossman. Written by Marti Noxon (Part I) and David Fury (Part II). Air date: 10/2/01.

"Afterlife." Dir. David Solomon. Written by Jane Espenson. Air date: 10/9/01.

"Life Serial." Dir. Nick Marck. Written by David Fury and Jane Espenson. Air date: 10/23/01.

"Smashed." Dir. Turi Meyer. Written by Drew Greenberg. Air date: 11/20/01.

"Gone." Dir. David Fury. Written by David Fury. Air date: 1/8/02.

"Hell's Bells." Dir. David Solomon. Written by Rebecca Rand Kirshner. Air date: 3/5/02.

"Normal Again." Dir. Rick Rosenthal. Written by Diego Gutierrez. Air date: 3/12/02.

"Seeing Red." Dir. Michael Gershman. Written by Steven DeKnight. Air date: 5/7/02.

"Villains." Dir. David Solomon. Written by Marti Noxon. Air date: 5/14/02.

"Two to Go." Dir. Bill Norton. Written by Douglas Petrie. Air date: 5/21/02.

"Grave." Dir. James Contner. Written by David Fury. Air date: 5/21/02.

Season 7

"Lessons." Dir. David Solomon. Written by Joss Whedon. Air date: 9/24/02.

"Same Time, Same Place." Dir. James Contner. Written by Jane Espenson. Air date: 10/8/02.

"Selfless." Dir. David Solomon. Written by Drew Goddard. Air date: 10/22/02.

"Conversations with Dead People." Dir. Nick Marck. Written by Jane Espenson and Drew Goddard. Air date: 11/5/02.

"Never Leave Me." Dir. David Solomon. Written by Drew Goddard. Air date: 11/26/02.

"Bring On the Night." Dir. David Grossman. Written by Marti Noxon and Douglas Petrie. Air date: 12/17/02.

"Showtime." Dir. Michael Grossman. Written by David Fury. Air date: 1/7/03.

"The Killer in Me." Dir. David Solomon. Written by Drew Greenberg. Air date: 2/4/03.

"Get It Done." Dir. Douglas Petrie. Written by Douglas Petrie. Air date: 2/18/03.

"Storyteller." Dir. Marita Grabiak. Written by Jane Espenson. Air date: 2/25/03.

"Dirty Girls." Dir. Michael Gershman. Written by Drew Goddard. Air date: 4/15/03.

"Empty Places." Dir. James Contner. Written by Drew Greenberg. Air date: 4/29/03.

"Touched." Dir. David Solomon. Written by Rebecca Rand Kirshner. Air date: 5/6/03.

"End of Days." Dir. Marita Grabiak. Written by Douglas Petrie and Jane Espenson. Air date: 5/13/03.

"Chosen." Dir. Joss Whedon. Written by Joss Whedon. Air date: 5/20/03.

ANGEL

Season 1

"City Of." Dir. Joss Whedon. Written by David Greenwalt and Joss Whedon. Air date: 10/5/99.

"Rm w/Vu." Dir. Scott McGinnis. Teleplay by Jane Espenson. Story by David Greenwalt and Jane Espenson. Air date: 11/2/99.

"I Will Remember You." Dir. David Grossman. Written by David Greenwalt and Jeannine Renshaw. Air date: 11/23/99.

"She." Dir. David Greenwalt. Written by Marti Noxon. Air date: 2/8/00.

"To Shanshu in L.A." Dir. David Greenwalt. Written by David Greenwalt. Air date: 5/23/00.

Season 2

"Judgment." Dir. Michael Lange. Teleplay by David Greenwalt. Story by Joss Whedon and David Greenwalt. Air date: 9/26/00.

"Are You Now or Have You Ever Been?" Dir. David Semel. Written by Tim Minear. Air date: 10/3/00.

"Untouched." Dir. Joss Whedon. Written by Mere Smith. Air date: 10/17/00.

"Darla." Dir. Tim Minear. Written by Tim Minear. Air date: 11/14/00.

"The Trial." Dir. Bruce Seth Green. Teleplay by Douglas Petrie and Tim Minear. Story by David Greenwalt. Air date: 11/28/00.

"Happy Anniversary." Dir. Bill Norton. Teleplay by David Greenwalt. Story by Joss Whedon and David Greenwalt. Air date: 2/6/01.

"The Thin Dead Line." Dir. Scott McGinnis. Written by Jim Kouf and Shawn Ryan. Air date: 2/13/01.

"Dead End." Dir. James Contner. Written by David Greenwalt. Air date: 4/24/01.

"Belonging." Dir. Turi Meyer. Written by Shawn Ryan. Air date: 5/1/01.

"Over the Rainbow." Dir. Fred Keller. Written by Mere Smith. Air date: 5/8/01.

"Through the Looking Glass." Dir. Tim Minear. Written by Tim Minear. Air date: 5/15/01.

"There's No Place Like Plrtz Glrb." Dir. David Greenwalt. Written by David Greenwalt. Air date: 5/22/01.

Season 3

“Heartthrob.” Dir. David Greenwalt. Written by David Greenwalt. Air date: 9/24/01.

“That Vision Thing.” Dir. Bill Norton. Written by Jeffrey Bell. Air date: 10/1/01.

“Offspring.” Dir. Turi Meyer. Written by David Greenwalt. Air date: 11/5/01.

“Quickening.” Dir. Skip Schoolnik. Written by Jeffrey Bell. Air date: 11/12/01.

“Lullaby.” Dir. Tim Minear. Written by Tim Minear. Air date: 11/19/01.

“Birthday.” Dir. Michael Grossman. Written by Mere Smith. Air date: 1/13/02.

“Waiting in the Wings.” Dir. Joss Whedon. Written by Joss Whedon. Air date: 2/4/02.

“Double or Nothing.” Dir. David Grossman. Written by David Goodman. Air date: 4/22/02.

“The Price.” Dir. Marita Grabiak. Written by David Fury. Air date: 4/29/02.

“A New World.” Dir. Tim Minear. Written by Jeffrey Bell. Air date: 5/6/02.

“Tomorrow.” Dir. David Greenwalt. Written by David Greenwalt. Air date: 5/20/02.

Season 4

"Deep Down." Dir. Terrence O'Hara. Written by Steven DeKnight. Air date: 10/6/02.

"Ground State." Dir. Michael Grossman. Written by Mere Smith. Air date: 10/13/02.

"The House Always Wins." Dir. Marita Grabiak. Written by David Fury. Air date: 10/20/02.

"Supersymmetry." Dir. Bill Norton. Written by Elizabeth Craft and Sarah Fain. Air date: 11/3/02.

"Apocalypse, Nowish." Dir. Vern Gillum. Written by Steven DeKnight. Air date: 11/17/02.

"Habeas Corpses." Dir. Skip Schoolnik. Written by Jeffrey Bell. Air date: 1/15/03.

"Long Day's Journey." Dir. Terrence O'Hara. Written by Mere Smith. Air date: 1/22/03.

"Salvage." Dir. Jefferson Kibbee. Written by David Fury. Air date: 2/19/03.

"Players." Dir. Michael Grossman. Written by Jeffrey Bell, Sarah Fain, and Elizabeth Craft. Air date: 3/26/03.

"Inside Out." Dir. Steven DeKnight. Written by Steven DeKnight. Air date: 4/2/03.

"Shiny Happy People." Dir. Marita Grabiak. Written by Elizabeth Craft and Sarah Fain. Air date: 4/9/03.

"The Magic Bullet." Dir. Jeffrey Bell. Written by Jeffrey Bell. Air date: 4/16/03.

"Sacrifice." Dir. David Straiton. Written by Ben Edlund. Air date: 4/23/03.

"Peace Out." Dir. Jefferson Kibbee. Written by David Fury. Air date: 4/30/03.

"Home." Dir. Tim Minear. Written by Tim Minear. Air date: 5/7/03.

Season 5

"Conviction." Dir. Joss Whedon. Written by Joss Whedon. Air date: 10/1/03.

"Just Rewards." Dir. James Contner. Teleplay by David Fury and Ben Edlund. Story by David Fury. Air date: 10/8/03.

"Hellbound." Dir. Steven DeKnight. Written by Steven DeKnight. Air date: 10/22/03.

"Lineage." Dir. Jefferson Kibbee. Written by Drew Goddard. Air date: 11/12/03.

"Destiny." Dir. Skip Schoolnik. Written by David Fury and Steven DeKnight. Air date: 11/19/03.

"Damage." Dir. Jefferson Kibbee. Written by Steven DeKnight and Drew Goddard. Air date: 1/28/04.

"Smile Time." Dir. Ben Edlund. Teleplay by Ben Edlund. Story by Joss Whedon and Ben Edlund. Air date: 2/18/04.

"A Hole in the World." Dir. Joss Whedon. Written by Joss Whedon. Air date: 2/24/04.

"Shells." Dir. Steven DeKnight. Written by Steven DeKnight. Air date: 3/3/04.

"Underneath." Dir. Skip Schoolnik. Written by Sarah Fain and Elizabeth Craft. Air date: 4/14/04.

"Origin." Dir. Terrence O'Hara. Written by Drew Goddard. Air date: 4/21/04.

"Time Bomb." Dir. Vern Gillum. Written by Ben Edlund. Air date: 4/28/04.

"Power Play." Dir. James Contner. Written by David Fury. Air date: 5/12/04.

"Not Fade Away." Dir. Jeffrey Bell. Written by Jeffrey Bell and Joss Whedon. Air date: 5/19/04.

GENERAL RESOURCES

Bails, Jennifer. "Pitt Scientists Resurrect Hope of Cheating Death," *Pittsburgh Tribune-Review*, June 29, 2005.

Ball, Philip. "Engineers Devise Invisibility Shield; Electron Effects Could Stop Objects from Scattering Light," *Science News*, February 28, 2005.

Bazhenov, V. A. *Piezoelectric Properties of Wood.* New York: Consultants Bureau, 1961.

Bloomfield, Louis. *How Things Work: The Physics of Everyday Life.* New York: John Wiley & Sons, 1977.

Bodanis, David. *$E = mc^2$: A Biography of the World's Most Famous Equation.* New York: Walker & Company, 2000.

Boyle, Alan. "Big Debate over a 'Little Big Bang': Real-Life Quest for Quarks Sparks Science-Fiction Speculation," MSNBC.com, September 21, 1999.

Britt, Robert Roy. "The Big Rip: New Theory Ends Universe by Shredding Everything," Space.com, March 6, 2003.

———. "New Theory: How to Make Objects Invisible," *LiveScience*, February 28, 2005.

Browne, Malcolm. "A Bet on a Cosmic Scale, and a Concession, Sort Of," *New York Times*, February 12, 1997.

Bruce, Colin. *Schrödinger's Rabbits: The Many Worlds of Quantum*. Washington, DC: Joseph Henry Press, 2004.

Chananie, Jon. "The Physics of Karate Strikes," *Journal of How Things Work*, fall 1999.

Cho, Adrian. "Constructing Spacetime—No Strings Attached," *Science*, November 8, 2002.

Clarke, Arthur C. *Profiles of the Future: An Inquiry into the Limits of the Possible*. New York: Warner Books, 1985.

Cole, K. C. *The Hole in the Universe: How Scientists Peered over the Edge of Emptiness and Found Everything*. New York: Harcourt, 2001.

Costa, M., with A. Goldberger and C. Peng. "Broken Asymmetry of the Human Heartbeat: Loss of Time Irreversibility in Aging and Disease," *Physical Review Letters*, November 4, 2005.

Coventry, Peter, with Roger Highfield. *The Arrow of Time: A Voyage Through Science to Solve Time's Greatest Mystery*. New York: Ballantine Books, 1992.

Davies, Paul. *How to Build a Time Machine.* New York: Penguin, 2001.

Davis, Wade. *The Serpent and the Rainbow.* New York: Warner Books, 1985.

Dennett, Daniel C. *Consciousness Explained.* New York: Gardners Books, 2004.

Faiola, Anthony. "Humanoids with Attitude," *Washington Post,* March 11, 2005.

Farivar, Cyrus. "Jeepers Creepers, Bionic Peepers," *Wired News,* May 5, 2005.

Ferguson, Kitty. *Measuring the Universe: Our Historic Quest to Chart the Horizons of Space and Time.* New York: Walker & Company, 1999.

Folger, Tim. "If an Electron Can Be in Two Places at Once, Why Can't You?," *Discover,* June 2005.

Fudge, Douglas, et al. "Composition, Morphology and Mechanics of Hagfish Slime," *The Journal of Experimental Biology,* 208, 4613–4625, December 2, 2005.

Gefter, Amanda. "Throwing Einstein for a Loop," *Scientific American,* December 2002.

Gilmore, Robert. *Alice in Quantumland.* New York: Springer-Verlag, 1995.

Golden, Christopher, with Stephen R. Bissette and Thomas E. Sniegoski. *The Monster Book.* New York: Simon & Schuster/Simon Pulse, 2002.

Gonick, Larry, with Art Huffman. *The Cartoon Guide to Physics.* Harper Collins, 1990.

Gott, J. Richard. *Time Travel in Einstein's Universe*. New York: Houghton Mifflin/Mariner Books, 2002.

Greene, Brian. *The Elegant Universe*. New York: Random House/Vintage, 2000.

———. *The Fabric of the Cosmos*. New York: Alfred A. Knopf, 2004.

Gribbin, John. *In Search of Schrödinger's Cat*. New York: Bantam, 1984.

Gugliotta, Guy. "A Challenge to Evolution of the Universe," *Washington Post*, January 12, 2006.

Hawking, Stephen. *A Brief History of Time*. New York: Bantam Books, 1988.

———. *Black Holes and Baby Universes*. New York: Bantam Books, 1994.

———. *The Universe in a Nutshell*. New York: Bantam Books, 2001.

Hawking, Stephen et al. *The Future of Space-Time*. New York: W. W. Norton, 2002.

Hecht, Jeff. "Quantum Teleporter Creates Laser Beam Clones," *New Scientist*, February 21, 2006.

Hogan, Jenny. "Ripples of Light Might Make You Invisible," *Nature*, March 5, 2005.

Holder, Nancy, with Jeff Mariotte and Maryelizabeth Hart. *Angel: The Casefiles, Vol. I*. New York: Simon & Schuster/Simon Pulse, 2002.

Holt, Jim. "My So-Called Universe," *Slate*, August 20, 2003.

Julsgaard, Brian, et al. "Experimental Long-Lived Entanglement of Two Macroscopic Objects," *Nature*, September 27, 2001.

Kaku, Michio. *Hyperspace: A Scientific Odyssey Through Parallel Universes, Time Warps, and the Tenth Dimension*. New York: Anchor Books, 1995.

———. *Parallel Worlds*. New York: Doubleday, 2005.

———. "Testing String Theory," *Discover*, August 2005.

Knight, Will. "U.S. and Canadian Skiers Get Smart Armour," *New Scientist*, February 14, 2006.

Krauss, Lawrence. *Beyond Star Trek*. New York: Harper Collins/Basic Books, 1997.

———. *Hiding in the Mirror: The Mysterious Allure of Extra Dimensions from Plato to String Theory and Beyond*. New York: Viking/Penguin, 2005.

———. *The Physics of Star Trek*. New York: Harper Collins/Basic Books, 1995.

Lane, Earl. "Story Is Out of This World," *Newsday*, July 21, 1999.

Lindley, David. *Where Does the Weirdness Go? Why Quantum Mechanics Is Strange, but Not as Strange as You Think*. New York: Harper-Collins/Basic Books, 1996.

Lockwood, Michael. *The Labyrinth of Time: Introducing the Universe*. Oxford: Oxford University Press, April 2005.

MacKay, Charles. *Extraordinary Popular Delusions and the Madness of Crowds*. New York: Three Rivers Press, 1980.

McKee, Maggie. "Strange New Object Found at Edge of Solar System," *New Scientist*, December 13, 2005.

———. "Wormhole Wanderers Face a Deadly Dilemma," *New Scientist*, May 24, 2005.

Mermin, N. David. *It's About Time: Understanding Einstein's Relativity*. Princeton: Princeton University Press, 2005.

Mone, Gregory. "The Fallacy of the Black Hole in Switzerland That Would Swallow the Earth," *Popular Science*, January 2004.

Morton, Oliver. "Deep Impact: It Came, It Seems, from Outer Space—and It Did So Quickly," Wired.com, February 2003.

Olsen, Stefanie. "Physicists Re-create Nature's Best Sound System," *CNET News*, June 22, 2005.

Overbye, Dennis. "Quantum Trickery: Testing Einstein's Strangest Theory," *New York Times*, December 27, 2005.

Owen, James. "Invisibility Shields Planned by Engineers," *National Geographic News*, February 28, 2005.

Penrose, Roger. *The Emperor's New Mind: Concerning Computers, Minds, and the Laws of Physics.* Oxford: Oxford University Press, 1990.

Peplow, Mark. "Gravity Doughnut Promises Time Machine," *Nature*, July 13, 2005.

Peterson, Ivars. "Surfing a Laser Wave: Toward a Tabletop Particle Accelerator; New Powerful Laser Pulses Generated," *Science News*, February 10, 1996.

———. "Timely Questions," *Science News*, March 28, 1992.

Randall, Lisa. *Warped Passages: Unraveling the Mysteries of the Universe's Hidden Dimensions.* New York: HarperCollins, 2005.

Reich, Eugenie Samuel. "Black Hole–Like Phenomena Created by Collider," *New Scientist*, March 16, 2005.

Rist, Curtis. "The Physics of Karate," *Discover*, May 2000.

Rothman, Tony. *Instant Physics: From Aristotle to Einstein, and Beyond*. New York: Fawcett Books, 1995.

Rovelli, Carlo. "Loop Quantum Gravity," *Physics World*, November 2003.

Ruditis, Paul and Diana G. Gallagher. *Angel: The Casefiles, Vol. II*. New York: Simon & Schuster/Simon Spotlight Entertainment, 2004.

Sandhana, Lakshmi. "Bionic Eyes Benefit the Blind," *Wired News*, July 16, 2003.

———. "Chips Coming to a Brain Near You," *Wired News*, October 22, 2004.

Schlermeler, Quirin. "The Long-Distance Thinker," *Nature*, January 6, 2005.

Searle, John R. *Minds, Brains, and Action*. Cambridge, MA: Harvard University Press, 1984.

Seife, Charles. *Alpha and Omega*. New York: Viking/Penguin, 2003.

Simon, J. Z. "The Physics of Time Travel," *Physics World*, December 1994.

Stafford, Nikki: *Bite Me: An Unofficial Guide to the World of Buffy the Vampire Slayer*. Toronto, Ontario (Canada): ECW Press, 2002.

———. *Once Bitten: An Unofficial Guide to the World of Angel*. Toronto, Ontario (Canada): ECW Press, 2004.

Stenger, Victor J. "Quantum Time Travel," *Skeptical Briefs*, June 2000.

Summers, Adam. "Slime and the Cytoskeleton," *Natural History Magazine*, October 2004.

Tegmark, Max. "Not Just a Figment of Science Fiction, Other Universes Are a Direct Implication of Cosmological Observations," *Scientific American*, May 2003.

Thorne, Kip. *Black Holes and Time Warps*. New York: W. W. Norton, 1995.

Tretkoff, Ernie. "RHIC Detects Liquid State of Quark-Gluon Matter," *APS News*, June 2005.

Trivedi, Bijal P. "Praying Mantis Uses Ultrasonic Hearing to Dodge Bats," *National Geographic Today*, November 19, 2002.

Walker, Jearle. "Karate Strikes," *American Journal of Physics* 43, (845–849), 1975.

———. "The Physics of Forces in Aikido," *Scientific American*, July 1980.

Warwick, Kevin. *I, Cyborg*. Urbana/Chicago: University of Illinois Press, 2004.

Weiss, Peter. "Collider Can't Cause Cosmic Calamity," *Science News*, October 23, 1999.

———. "Ion Collider, Doomsday Fears Rev Up," *Science News*, August 7, 1999.

———. "Realistic Time Machine? New Design Could Forego Exotic Ingredient," *Science News*, July 16, 2005.

———. "Seeking the Mother of All Matter: World's Mightiest Particle Collider May Transform Less-Than-Nothing into a Primordial Something," *Science News*, August 26, 2000.

Wheeler, John. "Of Wormholes, Time Machines and Paradoxes," *Astronomy*, February 1996.

Wilk, S. R. et al. "The Physics of Karate," *American Journal of Physics* 51, (783–790), 1983.

Witten, Edward. "Universe on a String," *Astronomy*, June 2002.

———. "When Symmetry Breaks Down," *Nature*, June 3, 2004.

ONLINE RESOURCES

Adams, Cecil. The Straight Dope: www.straightdope.com

Animal Planet: www.animal.discovery.com

Buffy vs. Angel (episode transcripts): www.buffy-vs-angel.com

How Stuff Works: www.howstuffworks.com

Oceanlink (hagfish): http://oceanlink.island.net/oinfo/hagfish/hagfishathome.html

Official String Theory Web Site: www.superstringtheory.com

Open Questions (M Theory): www.openquestions.com/oq-ph014.htm

Stakes and Salvation (Buffyverse encyclopedia): www.stakesandsalvation.com

Walker, Jearl. "The Physics of Forces in Judo," www.fightingarts.com/reading/article.php?id=281

Whedonesque (Joss Whedon's Web blog): www.whedonesque.com

Wikipedia: www.wikipedia.com

ACKNOWLEDGMENTS

It's a rare privilege when a writer has the opportunity to combine two very different passions into one pet project: in this case, physics and the Buffyverse. So I am grateful to my erstwhile Penguin editor, Caroline White, for enabling me to do just that. Caroline ignited my creative spark over lunch one day in New York City, during a broad discussion that encompassed both the difficulty of communicating the concepts of relativity to nonscientists, and our shared sorrow at the demise of one of our favorite TV series. I realized that it was possible to explain relativity and many other physics concepts using examples from *Buffy the Vampire Slayer* and *Angel*. And *The Physics of the Buffyverse* was born.

I feel equally privileged that Karen Anderson took over as the book's editor when Caroline departed, because she made it such a seamless transition. Karen capably steered the book through the long and complicated production process, providing invaluable guidance, reassurance, and good cheer along the way.

While writing the book, I received much-appreciated assistance from several scientists and science writers, who graciously overlooked the seeming silliness of the subject matter to offer their expertise. Gavin Braithwaite, Kenneth Chang, Alan Chodos, Jessica Clark, Moshe Gai, David Harris, Ted Hodapp, Diandra Leslie-Pelecky, Steve Pierson, James Riordon, Marc Sher, Francis Slakey, and Ben Stein all gave helpful comments (and/or performed the odd calculation) that kept the science on track. Extra special thanks go to materials physicist/fellow *Buffy* fan Brymer Chin for tipping me off to the piezoelectric properties of wood; providing insight into the physics of snot; and spending a chunk of his Thanksgiving weekend "investigating" the viscoelastic properties of raw cookie dough. That's what I call "curiosity-driven research." I thank all of the above for their willingness to indulge their sense of play, and hope they won't be too horrified by my creative application of their thoughtful input. Naturally, I shoulder the blame for the odd bit of conjecture and for any remaining technical errors.

Paul Dlugokencky deserves buckets of kudos for his wonderfully whimsical illustrations. The Roomful of Writers group—notably Jennifer Cohen Oko, Martha Heil, Elaine Heinzman, Pete Reppert, Kevin Ricche, Contessa Riggs, and the aforementioned James Riordon—offered their usual much-appreciated critiques. Their willingness to be brutally honest when something just wasn't working resulted in a much better book, even if it pinched a bit at times. Ann Kottner doggedly slogged through the entire draft manuscript, suggesting ways to improve overall clarity and continuity. Thanks are also due to my agent, Mildred Marmur, for her practical advice on the business end of book publishing (of which I remain woefully ignorant). And I am grateful to Dana Rosen in Penguin's legal department, for undertaking the thankless task of vetting every aspect of the book prior to publication. The prose is that much tighter because of her efforts.

Just as I was finishing the final revisions to the manuscript, I (belatedly) discovered the wonderful world of the blogosphere. Not only is it a natural outlet for my writerly effluvia, it is chock-full of smart, witty, science-and-art-minded folk, many of whom turned out to be kindred spirits. Abbas Raza, Robin Varghese, and Morgan Meis at 3 Quarks Daily quickly became friends. So did Sean Carroll of Cosmic Variance, who helped hone my poker skills and refine my thinking about the intricate interplay among science, myth, and metaphor. I wish I'd had the benefit of his companionship and keen insights sooner.

I am, as always, both warmed and humbled by the unwavering love, support, and infinite patience shown to me by friends and family. A few deserve special mention. Bob Mondello and Carlos Schröder saved my sanity during the most intensive periods of writing by insisting I leave the computer on occasion to come over for a home-cooked meal and actual social interaction. Peri Lyons Thalenberg boosted my flagging spirits time and again with her rapier wit, soulful wisdom, and relentless *bonhomie*. Rich Kim (aka "the human Google") looked up bits of random trivia, shared his martial arts expertise, and served as a valuable sounding board as I was developing my ideas. He also makes a mean bowl of *duk guk*. And what can I say about my parents, Paul and Jeanne, who remain my number-one fans? I love you both.

Last, but far from least, I must thank series creator Joss Whedon and his fabulous team of writers for creating such a richly textured fictional universe. I am not the first to be inspired by his prolific imagination, and I most certainly will not be the last.

INDEX

A

B

C

D

E

F

G

H

I

J

K

L

M

N

O

P

Q

R

S

T

U

V

W

X

Y

Z